D0857839

A Scrapbook of
Complex Curve Theory

THE UNIVERSITY SERIES IN MATHEMATICS

Series Editor: Joseph J. Kohn
  Princeton University

**INTRODUCTION TO PSEUDODIFFERENTIAL
AND FOURIER INTEGRAL OPERATORS**
François Treves
VOLUME 1: PSEUDODIFFERENTIAL OPERATORS
VOLUME 2: FOURIER INTEGRAL OPERATORS

A SCRAPBOOK OF COMPLEX CURVE THEORY
C. Herbert Clemens

A Scrapbook of
Complex Curve Theory

C. Herbert Clemens

University of Utah
Salt Lake City, Utah

PLENUM PRESS · NEW YORK AND LONDON

Library of Congress Cataloging in Publication Data

Clemens, Charles Herbert, 1939-
  A Scrapbook of complex curve theory.

  (The University series in mathematics)
  Bibliography: p.
  Includes index.
  1. Curves, Algebraic. 2. Functions, Theta. 3. Jacobi varieties. I. Title. II. Series:
University series in mathematics (New York, 1980-   )
QA565.C55   516.3´5   80-20214
ISBN 0-306-40536-9

Printed in the United States of America

Preface

This is a book of "impressions" of a journey through the theory of complex algebraic curves. It is neither self-contained, balanced, nor particularly tightly organized. As with any notebook made on a journey, what appears is that which strikes the writer's fancy. Some topics appear because of their compelling intrinsic beauty. Others are left out because, for all their importance, the traveler found them boring or was too dull or lazy to give them their due.

Looking back at the end of the journey, one can see that a common theme in fact does emerge, as is so often the case; that theme is the theory of theta functions. In fact very much of the material in the book is preparation for our study of the final topic, the so-called Schottky problem. More than once, in fact, we tear ourselves away from interesting topics leading elsewhere and return to our main route.

Some of the subjects are extremely elementary. In fact, we begin with some musings in the vicinity of secondary-school algebra. Later, on occasion and without much warning, we jump into some fairly deep water. Our intent is to struggle with some deep topics in much the same way that a beginning researcher might, using whatever tools we have at hand or can grab somehow or other. Sometimes we use no background material and do everything in detail; sometimes we use some of the heaviest of modern machinery. We hope to motivate further study or, preferably, further discussion with an expert in the field. In short, our aim is to motivate and stimulate mathematical activity rather than to present a finished product, and our point of view is romantic rather than rigorous.

The material treated here was originally brought together for a Summer Course of the Italian National Research Council held in Cortona, Italy, in 1976. It comes from so many sources that adequate acknowledgment would be difficult. The treatment of real two-dimensional geometries of constant curvature comes from Cartan's classic text on Riemannian geometry; several items concerning the arithmetic of curves are borrowed from Serre's lovely book, *A Course in Arithmetic*; Manin's beautiful theorem on rational points of elliptic curves given in Chapter Two was explained to the author by A. Beauville; some of the theta identities in Chapter Three are lifted from the famous analysis text of Whittaker and Watson; and the construction of the level-two moduli space for elliptic curves was motivated by David Mumford's way of viewing the moduli space of curves of a fixed genus. The discussion of the Jacobian variety in Chapter Four leans heavily on work of Joseph Lewittes, and the discussion of the Schottky problem comes from work of Accola, Farkas, Igusa, and Rauch. But perhaps the author's greatest debt is to Phillip Griffiths, through whom he came to enjoy this subject.

The author also wishes to thank Sylvia M. Morris, Mathematics Department of the University of Utah, for preparing the manuscript, and Toni W. Bunker, of the same department, for preparing the figures.

Herbert Clemens
Salt Lake City, Utah

Contents

Notation

Most of the notation used in this book is quite standard, for example,

$\mathbb{Z}$ = ring of integers,

$\mathbb{Q}$ = field of rational numbers,

$\mathbb{R}$ = field of real numbers,

$\mathbb{C}$ = field of complex numbers.

Each of the six chapters is divided into sections, for instance, Chapter Three has Sections 3.1, 3.2, etc. Equations are numbered consecutively within chapters—(3.1), (3.2), etc.—as are the figures.

Square brackets will be used to enclose matrices and are also used later in the book in expressions involving theta functions with characteristic, for example, $\theta[{}^1_1](u; \tau)$

When there are complicated exponents, the exp form of the exponential is used with the convention $\exp\{x\} = e^x$.

Conics

1.1 Hyperbola Shadows

Let's start our study of curves in a very elementary way, with the curves that we learned about in school, and let's engage in some reasoning about these curves just as we might have done when we first met them. This is all very elementary, of course, but it will set the proper tone for what follows—much less elementary material approached in a similar way.

Sometime or other during his formal schooling, every student of mathematics has studied the set of solutions to the equation

$$Ax^2 + Bxy + Cy^2 + Dx + Ey + F = 0, \tag{1.1}$$

where A, B, C, D, E, and F are given constants. If these constants are real, then the set of solutions to equation (1.1) may form

an ellipse	two parallel lines	a point	
a hyperbola	two crossing lines	the empty set	(1.2)
a parabola	one line "counted twice"	the whole plane	

in the real plane $\mathbb{R} \times \mathbb{R}$. The notions of ellipse, hyperbola, and parabola can be unified. To do this, we place the (x, y) plane in three-dimensional space by considering it to be the set of points

$$\{(x, y, z): z = 1\}$$

in $\mathbb{R}^3$ (Figure 1.1). Suppose we start, say, with the hyperbola

$$\begin{aligned} xy &= 1, \\ z &= 1 \end{aligned} \tag{1.3}$$

and (Figure 1.2) put a light at $(0, 0, 0)$. To see the shadow of our hyperbola, we take a movable plane, which we will call a movie screen, and place

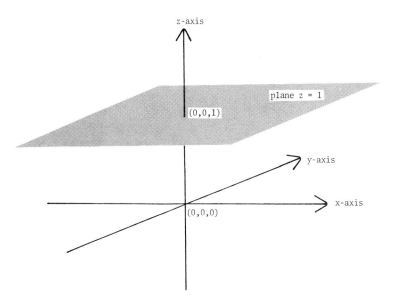

Figure 1.1. Putting the (x, y) plane into $\mathbb{R}^3$.

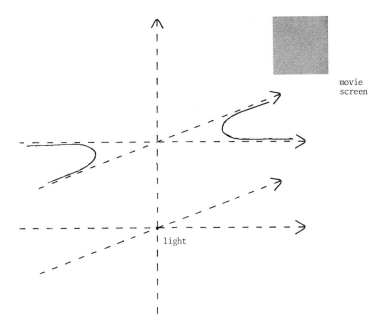

Figure 1.2. Projecting from the origin.

it in various positions in space so that we can see the shadow of our hyperbola from many different angles. If our movie screen is the plane

$$z = 2,$$

then we see a shadow which is simply a bigger hyperbola, but if our movie screen is the plane

$$x = 2,$$

then what we see is no longer a hyperbola but a piece of a parabola (Figure 1.3). Let's check this computationally. Any point of the shadow on the movie screen $x = 2$ must lie on some line passing through $(0, 0, 0)$ and a point of the hyperbola given in equation (1.3). So for each such point (x, y, z) on the shadow, there must exist a real number λ such that

$$\begin{aligned} \lambda z &= 1, \\ (\lambda x)(\lambda y) &= 1. \end{aligned} \tag{1.4}$$

In other words,

$$\frac{1}{z^2} xy = 1,$$

or

$$xy = z^2 \qquad \text{and} \qquad z \neq 0.$$

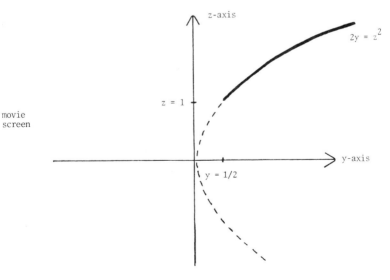

Figure 1.3. The shadow is now a parabola.

But now if such a point (x, y, z) lies in the movie screen $x = 2$, it must lie on the parabola

$$2y = z^2.$$

Finally, we are talking about a *shadow*, so the number λ in equation (1.4) must satisfy the inequality

$$0 < \lambda \leq 1,$$

that is,

$$z > 1.$$

In a similar way, if we put our movie screen in the position

$$x + y = 4,$$

we obtain as the shadow the set of points (x, y, z) such that

$$
\begin{aligned}
x + y &= 4, \\
xy &= z^2, \\
z &\geq 1.
\end{aligned}
\tag{1.5}
$$

This shadow is a piece of an ellipse. To see this, rotate three-dimensional space 45° around the z axis by means of the transformation

$$x \mapsto \frac{1}{\sqrt{2}}(x - y),$$

$$y \mapsto \frac{1}{\sqrt{2}}(x + y),$$

$$z \mapsto z.$$

The plane $x + y = 4$ has, as its preimage under this transformation, the plane

$$x = 2\sqrt{2}.$$

The set (1.5) has preimage

$$
\begin{aligned}
x &= 2\sqrt{2}, \\
\tfrac{1}{2}(x^2 - y^2) &= z^2, \\
z &\geq 1.
\end{aligned}
$$

In other words, in our rotated setup, our rotated hyperbola has the shadow

$$z^2 + \tfrac{1}{2}y^2 = 4,$$

$$z \geq 1$$

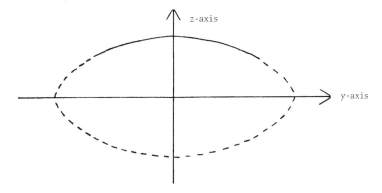

Figure 1.4. Now an elliptical shadow.

on our rotated movie screen $x = 2\sqrt{2}$. So the shadow is a piece of an ellipse (Figure 1.4).

1.2 Real Projective Space, The "Unifier"

Now it seems clear that we will have an easier time understanding the behavior of the various projections of our solution set onto the movable screen if we do not allow ourselves to be limited by the "extraneous" physical properties of our model.

In particular, things will be easier to grasp if we let the "shadow" consist of *all* points of intersection of the cone of solutions

$$xy = z^2$$

with our movable screen

$$ax + by + cz + d = 0.$$

In fact, if our line of thought is to equate any two solution sets

(a) which are "slices" of the same cone, and
(b) from which the original cone can be reconstructed,

then we might as well forget altogether about making slices and simply consider the cones themselves. That is, instead of studying the solution sets to equation (1.1) in $\mathbb{R}^2$, we will study solution sets to the *homogeneous* equation

$$Ax^2 + Bxy + Cy^2 + Dxz + Eyz + Fz^2 = 0 \qquad (1.6)$$

in $\mathbb{R}^3$. Written in (symmetric) matrix form, this last equation becomes

$$[x \quad y \quad z] \begin{bmatrix} A & B/2 & D/2 \\ B/2 & C & E/2 \\ D/2 & E/2 & F \end{bmatrix} \begin{bmatrix} x \\ y \\ z \end{bmatrix} = 0,$$

or simply

$$v \cdot M \cdot {}^t v = 0, \tag{1.7}$$

where $v = (x, y, z)$ is a vector in $\mathbb{R}^3$ and M is a symmetric 3×3 matrix.

Suppose now we make a linear change of the coordinates (x, y, z) as follows. We map $\mathbb{R}^3$ onto itself by the transformation

$$\begin{aligned} x &\mapsto (l_{11}x + l_{12}y + l_{13}z), \\ y &\mapsto (l_{21}x + l_{22}y + l_{23}z), \\ z &\mapsto (l_{31}x + l_{32}y + l_{33}z), \end{aligned} \tag{1.8}$$

where $L = (l_{ij})$ is an invertible 3×3 matrix. The cone (1.7) then has, as preimage under this mapping, the cone

$$v \cdot ({}^t LML) \cdot {}^t v = 0.$$

In order to break cones into types in the same spirit in which we prepared the list (1.2), we will say that two cones are equivalent if some equation for one of them can be obtained from some equation for the other by a linear change of coordinates (1.8).

Now put

$$(v_1, v_2) = v_1 \cdot M \cdot {}^t v_2, \tag{1.9}$$

where $v_i = (x_i, y_i, z_i) \in \mathbb{R}^3$. Either

$$A = B = \cdots = F = 0$$

or there is a vector $v_1 \in \mathbb{R}^3$ such that

$$(v_1, v_1) = \pm 1.$$

By a (linear) change of coordinates we can find an equivalent cone for which

$$v_1 = (1, 0, 0)$$

and for which the set of vectors "perpendicular" to v_1, that is,

$$\{v \in \mathbb{R}^3 : (v, v_1) = 0\},$$

is simply the (y, z) plane. The equation (1.7) of this new cone will be of the form

$$[x \; y \; z] \begin{bmatrix} \pm 1 & 0 & 0 \\ 0 & C & E/2 \\ 0 & E/2 & F \end{bmatrix} \begin{bmatrix} x \\ y \\ z \end{bmatrix} = 0. \qquad (1.10)$$

Again either $C = E = F = 0$ or there is a vector v_2 in the (y, z) plane such that for the cone (1.10),

$$(v_2, v_2) = \pm 1.$$

After a further change of coordinates we can then obtain an equivalent cone whose equation is of the form

$$[x \; y \; z] \begin{bmatrix} \varepsilon_1 & 0 & 0 \\ 0 & \varepsilon_2 & 0 \\ 0 & 0 & \varepsilon_3 \end{bmatrix} \begin{bmatrix} x \\ y \\ z \end{bmatrix} = 0, \qquad (1.11)$$

where $\varepsilon_1 \geq \varepsilon_2 \geq 0$, $\varepsilon_2 \geq \varepsilon_3$, and $\varepsilon_i \in \{0, \pm 1\}$. As shown in Table A, the list (1.2) of types of cones is considerably shortened.

1.3 Complex Projective Space, The Great "Unifier"

Now suppose we begin our entire discussion again. Let A, B, ..., $F \in \mathbb{C}$, the complex numbers, let our solution set lie in $\mathbb{C}^2$, etc. Again we pass to complex "cones," λ in equations (1.4) becomes a complex scalar, etc. Again we equate cones if they can be transformed, one into the other, by an invertible linear transformation of $\mathbb{C}^3$. The list of possibilities, given in Table B, is now even shorter (since every second-degree equation has complex solutions).

Table A

ε_1	ε_2	ε_3	Description
1	1	1	Point
1	1	0	Line
1	1	-1	"Usual" cone
1	0	0	Plane counted twice
1	0	-1	Two intersecting planes
0	0	0	$\mathbb{R}^3$

Table B

ε_1	ε_2	ε_3	Description
1	1	1	"Usual" cone
1	1	0	Two intersecting planes
1	0	0	One plane counted twice
0	0	0	$\mathbb{C}^3$

We will call the set of complex lines through the origin in $\mathbb{C}^3$ the

complex projective plane

and denote it by $\mathbb{C}\mathbb{P}_2$. It can be thought of as simply the usual complex plane with "ideal points" at infinity, one ideal point for each complex line through the origin. The relation between the usual (affine) coordinates (x, y) of the plane $\mathbb{C}^2$ and the (homogeneous) coordinates (x, y, z) of the space of lines through the origin in $\mathbb{C}^3$ is as follows: if the affine set of points is given by the equation

$$\sum_{m, n \leq N} A_{mn} x^m y^n = 0, \tag{1.12}$$

the projective set of complex lines is given by

$$\sum_{m, n \leq N} A_{mn} x^m y^n z^{(N - (m+n))} = 0. \tag{1.13}$$

We can freely pass from the affine object to its projectivization, which is well determined as long as there is some $A_{mn} \neq 0$ with $(m + n) = N$. Then all but a finite number $(\leq N)$ of elements (lines) of the projective object correspond to elements (points) in the affine object. Conversely, we pass from the projective object (1.13) to the affine object (1.12) by setting $z = 1$, that is, by intersecting the "cone" (1.13) with the complex plane $z = 1$ in $\mathbb{C}^3$. From now on it will be convenient to call the elements of $\mathbb{C}\mathbb{P}_2$ "points" rather than "lines."

So now let us look at the solution set for the homogeneous equation (1.6) in $\mathbb{C}\mathbb{P}_2$. First, we will always assume that $A, B, \ldots, F$ are not all 0. Also notice that if we multiply all six of these numbers by the same non-zero complex scalar, the solution set does not change. Thus the set of "conics" in $\mathbb{C}\mathbb{P}_2$, i.e., solution sets for equations of the form (1.6), has no more than five parameters; the conic is determined by the line through 0 and $(A, B, \ldots, F)$ in $\mathbb{C}^6$.

1.4 Linear Families of Conics

Let's move on now to a less elementary setting. We wish to consider the set of all conics and reflect on some of the geometric properties of this set. We will need a basic concept from algebra, the so-called *resultant* of a system of two equations. It is the device which tells us when the two equations have a common solution.

Suppose now that we consider two homogeneous polynomial equations

$$F(x, y, z) = \sum_{i=0}^{m} r_i(x, y)z^i = 0,$$

$$G(x, y, z) = \sum_{i=0}^{n} s_i(x, y)z^i = 0,$$
(1.14)

where r_i and s_i are homogeneous polynomials of degree $(m - i)$ and $(n - i)$ in (x, y). By elimination theory (van der Waerden [11], vol. 1, pp. 83–85), for a fixed value of (x, y) the resulting polynomial equations in z have a common solution if and only if

$$\det \begin{bmatrix} r_m & r_{m-1} & \cdots & & & r_0 & & \\ & r_m & r_{m-1} & \cdots & & & r_0 & \\ & & (n \text{ times}) & & & & & \\ & & & r_m & r_{m-1} & \cdots & r_0 \\ s_n & s_{n-1} & \cdots & & & s_0 & & \\ & s_n & s_{n-1} & \cdots & & & s_0 & \\ & & (m \text{ times}) & & & & & \\ & & & s_n & s_{n-1} & \cdots & s_0 \end{bmatrix} = 0.$$

This is a homogeneous equation of degree mn in (x, y). If it vanishes identically, then elimination theory tells us that the two polynomials F and G [considered to be elements of the ring of polynomials in z with coefficients in the quotient field of the ring of polynomials in (x, y)] have a common factor. Applying Gauss's lemma (van der Waerden [11], vol. 1, pp. 70–72), they must in fact have a common factor in the ring of homogeneous polynomials in (x, y, z). So if F and G are homogeneous of degree 2 and have the same solution set, we can conclude that

$$F = \lambda G, \qquad \lambda \neq 0.$$

Thus the set of "conics," that is, the set of solution sets to equations of the form (1.6), are parametrized by a complex projective five-space (that is, the set of all complex lines through the origin in $\mathbb{C}^6$) with homogeneous coordinates

$$(A, B, C, D, E, F). \tag{1.15}$$

So now we want to look at the geometry of the set of conics that fulfill some particular condition. For example, the set of conics passing through the point

$$(1, 0, 0)$$

is given by the equation

$$A = 0,$$

and so these conics form a complex projective four-space. What happens if we wish to study those conics passing through two or more points? Each new point imposes another linear condition on $(A. B, ..., F)$. However, we must look at the independence of these conditions. Are some of the conditions redundant? It is very easy to see that if we choose our points sufficiently generally, then each additional point imposes an independent condition until we run out of conics. But let's show this in a way that will work in the projective plane

$$\mathbb{FP}_2$$

over many fields $\mathbb{F}$ and for homogeneous forms of different degrees. We start with the set of monomials

$$x^i y^j z^k, \tag{1.16}$$

where $(i + j + k) =$ some fixed number N. If we make the substitution

$$y = x^r, \qquad z = x^s$$

for an appropriate fixed choice of r and s, then it will never be true that

$$i + jr + ks = i' + j'r + k's,$$

where $x^i y^j z^k$ and $x^{i'} y^{j'} z^{k'}$ are distinct monomials in the set (1.16). Recalling the theory of the van der Monde determinant (Lang [5], p. 179), we see that if $M = \max\{i + rj + sk: i + j + k = N\}$, then the matrix

$$[(x_l)^{i + rj + sk}], \qquad i + j + k = N, \, l = 0, ..., M,$$

must have maximal rank whenever the numbers

$$x_0, ..., x_M$$

are pairwise distinct. So choosing

$$y_l = x_l^r, \qquad z_l = x_l^s,$$

we find among the $(M + 1)$ points

$$(x_l, y_l, z_l) \in \mathbb{C}\mathbb{P}_2,$$

a subset whose cardinality is equal to that of the set (1.16) with the property that the square matrix

$$[(x_l^i y_l^j z_l^k)], \qquad i + j + k = N, \ l \in \text{subset}, \tag{1.17}$$

has nonzero determinant. Applying this reasoning to the set of monomials

$$x^2, \ xy, \ y^2, \ xz, \ yz, \ z^2,$$

we conclude that for the proper choice of six points in $\mathbb{C}\mathbb{P}_2$, there is no conic passing through them all, and for the proper choice of five points there is a unique conic containing them. In fact, since the condition that six points be contained in a conic is simply the polynomial condition that the determinant of the matrix (1.17) vanish, the set of sextuples of points which contain no conic is an open dense subset of

$$(\mathbb{C}\mathbb{P}_2)^6.$$

Similarly, the set of quintuples of points which determine a unique conic is an open dense subset of

$$(\mathbb{C}\mathbb{P}_2)^5.$$

1.5 The Mystic Hexagon

There is some beautiful classical geometry connected with these facts—for instance, the problem of how to construct the conic through five given points, which dates from ancient times. The most elegant solution, perhaps, is given by Pascal's "mystic hexagon." Namely, suppose we are given five points A, B, C, D, E in order. Referring to Figure 1.5, we make the following geometric construction:

- Construct the lines AB and DE to determine their point of intersection P_1.
- Draw any line through A and let P_2 denote its intersection with the line CD.
- Then, if P_3 is the point of intersection of the line BC with the line $P_1 P_2$, the point P of intersection of the lines AP_2 and EP_3 will lie on the conic $ABCDE$. That is because intercepts of opposite edges of a hexagon inscribed in a conic lie on a line.

To see why this last sentence is true, let Q denote the conic $ABCDE$. Now

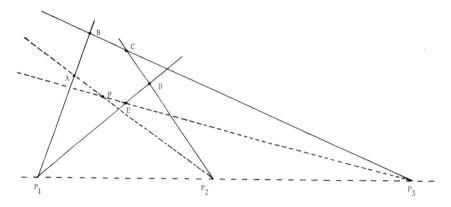

Figure 1.5. The point P traces out a conic through A, B, C, D, E.

consider the following list of three cubics [that is, solution sets to homogeneous polynomials of degree 3 in (x, y, z)]:

$$Q \text{ plus } P_1P_2,$$
$$AB \text{ plus } CD \text{ plus } EP, \qquad\qquad (1.18)$$
$$PA \text{ plus } BC \text{ plus } DE.$$

All three cubics contain the eight points

$$A, B, C, D, E, P_1, P_2, P_3. \qquad\qquad (1.19)$$

Now we have seen that "in general" we should expect the set of cubics containing eight points to be a $\mathbb{CP}_1$, that is, the set of homogeneous polynomials in (x, y, z) of degree 3 which vanish at eight general points is a two-dimensional vector space. It can be checked that if A, B, C, D, E are chosen in such a way that the conic is uniquely determined and not degenerate [that is, no $\varepsilon_i = 0$ in (1.11)], then the eight points in (1.19) are sufficiently general, so that the equations defining the three cubics in (1.18) must be linearly dependent. (See Griffiths and Harris [1], pp. 671–673.) Thus any common solution to two of them will also be a solution to the third. Since P is a common solution for the last two, it must also lie on the conic Q.

Another basic principle we've brushed up against here is *Bezout's theorem*, that if C_1 is a plane curve of degree m [solution set of a homogeneous polynomial of degree m in (x, y, z)] and C_2 is a plane curve of degree n, then either C_1 and C_2 have a component in common or they have no more than mn points in common. This is clear from our computation of the resultant of the system of equations (1.14). The resultant is a homogeneous polynomial of degree mn in (x, y).

1.6 The Cross Ratio

Another topic of interest in connection with the study of conics is the *cross ratio*. Let's approach this as it must have been approached originally. Figure 1.6, together with the identity

$$\sin \beta = \sin(\pi - \beta),$$

shows immediately that

$$(\sin \alpha)/A = (\sin \beta)/B.$$

From these last two identities and Figure 1.7 we obtain that

$$\frac{A/B}{A'/B'} = \frac{\sin \alpha/\sin \beta}{\sin \alpha'/\sin \beta'} = \frac{\sin \alpha}{\sin \alpha'},$$

and so we get the cross ratio

$$\frac{\overline{AC}/\overline{BC}}{\overline{AD}/\overline{BD}} = \frac{\sin \alpha/\sin(\beta + \gamma)}{\sin(\alpha + \beta)/\sin \gamma}$$

(see Figure 1.8). This expression is called the cross rat o of the points $(ABCD)$ and is clearly unchanged if the line L is moved. This leads us to the fact (which is easily checked by direct computation) that if

$$(x_1, y_1), \quad (x_2, y_2), \quad (x_3, y_3), \quad (x_4, y_4)$$

are four distinct points of $\mathbb{CP}_1$ and if

$$(x, y) \mapsto (ax + by, cx + dy) \tag{1.20}$$

is a linear transformation of $\mathbb{CP}_1$ onto itself, then the expression

$$\frac{(x_1 y_3 - y_1 y_3)/(x_2 y_3 - y_2 x_3)}{(x_1 y_4 - y_1 x_4)/(x_2 y_4 - y_2 x_4)}$$

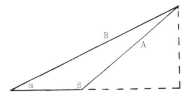

Figure 1.6. The law of the sine.

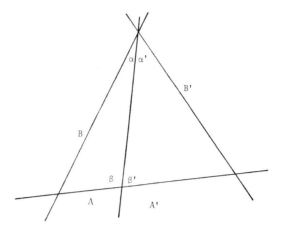

Figure 1.7. Ratio of sines of two angles, α and α', with a common vertex.

is a (nonzero) invariant under transformations of type (1.20) and is called, of course, the cross ratio

$$(P_1, P_2, P_3, P_4),$$

where $P_i = (x_i, y_i)$. Also notice that the cross ratio of four points on a conic is well defined by the fact that the transformation f of $\mathbb{CP}_1$ given by Figure 1.9 is linear. Since all nondegenerate conics are equivalent, it suffices to check this last statement for the conic given in affine coordinates by

$$x^2 + y^2 = 1,$$

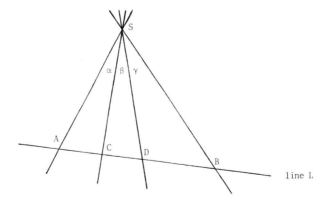

Figure 1.8. The cross ratio.

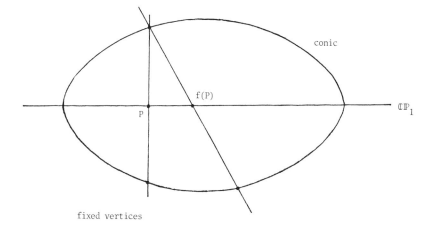

Figure 1.9. Projecting a conic from different vertices.

with fixed vertices (a, b) and $(a, -b)$ and $\mathbb{CP}_1$ represented by the x axis as in Figure 1.12. In this special case the statement is again a nice fact about plane geometry which we will now examine.

First, notice from Figure 1.10 that

$$2\alpha + \lambda = \pi,$$
$$2\beta + \mu = \pi,$$
$$2\gamma + \nu = \pi,$$

so that

$$\alpha + \beta + \gamma = \pi.$$

From this it follows that the triangles ACS and DBS are similar, and so

$$\overline{SA} \cdot \overline{SB} = \overline{SC} \cdot \overline{SD}.$$

Thus the product $\overline{SA} \cdot \overline{SB}$ is independent of the line through S and so the triangles SAC and SDB in Figure 1.11 are similar; therefore

$$\alpha = \delta.$$

Next suppose that in Figure 1.11 we arrange things so that

$$\overline{OD} \perp \overline{BC}.$$

Then

$$\alpha = \delta = \varepsilon/2.$$

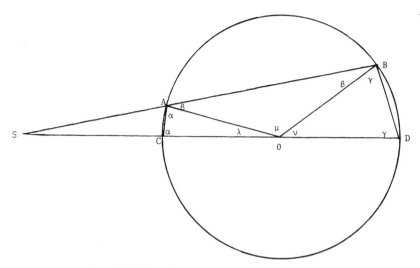

Figure 1.10. First figure on rays intersecting a circle.

Now we are ready to show why the mapping defined in Figure 1.9 is a linear automorphism of $\mathbb{C}\mathbb{P}_1$, that is, a transformation of type (1.20). We consider Figure 1.12. By what we did before we know that $\delta = \frac{1}{2}(2\alpha) = \alpha$ so that $\beta = \gamma = \delta + \varepsilon$. Thus the triangles ORB and OBS are similar, from which it follows that

$$\overline{OR} \cdot \overline{OS} = \overline{OB}^2.$$

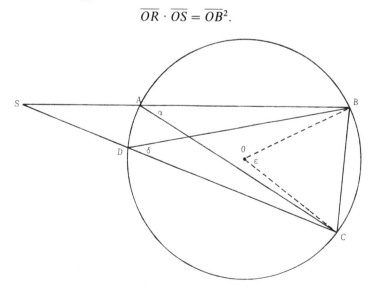

Figure 1.11. Second figure on rays intersecting a circle.

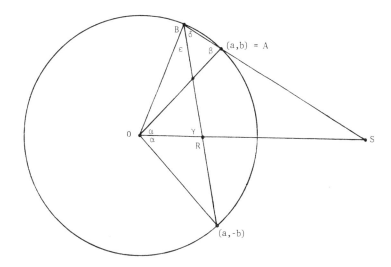

Figure 1.12. Projecting the circle onto a line from vertices (a, b) and $(a, -b)$.

This formula shows why the mapping f in Figure 1.9 is linear, since in the special case in which the conic is the unit circle the mapping is simply

$$x \mapsto 1/x.$$

1.7 Cayley's Way of Doing Geometries of Constant Curvature

Now let's follow an irresistible side topic related to conics but leading in an entirely different direction, that of Riemannian geometry. Riemannian geometry found its origins in efforts to show the independence of Euclid's fifth postulate—that, through a point not on a line, there passes a unique line which does not intersect the given one. Eventually two-dimensional worlds were produced which verified all Euclid's postulates except the fifth, and each of these "geometries" was uniform in the sense that it looked the same no matter at which point you stood in its two-dimensional world. The English mathematician Cayley found a lovely way to study all these geometries at once, and his main tools were the cross ratio and the theory of conics. In the next five sections we will touch on the high points of Cayley's theory. The beginning point is the solution set to the equation

$$x^2 + y^2 + z^2 = 0. \tag{1.21}$$

This conic in $\mathbb{C}\mathbb{P}_2$ will be called the *absolute*.

Given two points on the unit sphere

$$P_1, P_2 \in S = \{(x, y, z) \in \mathbb{R}^3 : x^2 + y^2 + z^2 = 1\},$$

the distance between them on S is the same as the angle δ between the lines OP_1 and OP_2 measured in radians. Write

$$P_i = (x_i, y_i, z_i)$$

and denote the complex projective line

$$\{P_1 + \lambda P_2 : \lambda \in (\mathbb{C} \cup \{\infty\})\} \subseteq \mathbb{CP}_2$$

by L. Let

$$\{R_1, R_2\} = L \cap Q,$$

where Q is the standard cone (1.21). These two points R_1 and R_2 correspond to values λ_1 and λ_2 of λ given by the roots of the equation

$$(x_1 + \lambda x_2)^2 + (y_1 + \lambda y_2)^2 + (z_1 + \lambda z_2)^2 = 0,$$

that is,

$$1 + (\cos \delta)2\lambda + \lambda^2 = 0,$$

and by the quadratic formula

$$\frac{\lambda_1}{\lambda_2} = \frac{\cos \delta \pm (\cos^2 \delta - 1)^{1/2}}{\cos \delta \mp (\cos^2 \delta - 1)^{1/2}} = \frac{\cos \delta \pm i \sin \delta}{\cos \delta \mp i \sin \delta} = e^{\pm 2i\delta}.$$

On the other hand, we compute the cross ratio

$$(P_1, P_2, R_1, R_2) = \lambda_1/\lambda_2.$$

Thus we have a nice distance formula on the two-sphere S,

$$\delta = \pm \frac{1}{2i} \log(P_1, P_2, R_1, R_2),$$

where R_1 and R_2 are the two points at which the line through P_1 and P_2 crosses the standard conic (1.21), the absolute.

Of course, we can project half the sphere S onto the plane

$$x, y \in \mathbb{R}, \qquad z = 1, \tag{1.22}$$

by central projection. This correspondence then gives a way of measuring distances between points on the plane (1.22) by measuring the angle between the corresponding points on S. But since the cross ratio is a projective invariant, we have a way to compute the distance which involves only the *complex* plane $z = 1$. Namely, let

$$p_1, p_2 \in \mathbb{R}^2,$$

and let r_1 and r_2 be the two points at which the *complex* line through p_1 and p_2 intersects the conic

$$x^2 + y^2 = -1 \tag{1.23}$$

in $\mathbb{C}^2$. Then the (spherical) distance between p_1 and p_2 is given by the formula

$$\pm \frac{1}{2i} \log(p_1, p_2, r_1, r_2). \tag{1.24}$$

This is because the cross ratio is preserved under central projection. From the invariance of the cross-ratio, we also conclude that distance is preserved by any transformation

$$
\begin{aligned}
x &\mapsto \frac{a_{11}x + a_{12}y + a_{13}}{a_{31}x + a_{32}y + a_{33}}, \\
y &\mapsto \frac{a_{21}x + a_{22}y + a_{23}}{a_{31}x + a_{32}y + a_{33}}
\end{aligned}
\tag{1.25}
$$

of $\mathbb{C}^2$ which is real (i.e., $a_{ij} \in \mathbb{R}$) and which preserves the conic (1.23). Of course, this is simply a translation of the condition that the matrix

$$[a_{ij}]$$

be orthogonal.

Now suppose we replace the sphere S by the surface S_K given by

$$K(x^2 + y^2) + z^2 = 1.$$

Suppose we measure the distance between two points (x_1, y_1, z_1) and (x_2, y_2, z_2) on S_K by measuring the distance between $(x_1, y_1, K^{-1/2}z_1)$ and $(x_2, y_2, K^{-1/2}z_2)$ on the sphere of radius $K^{-1/2}$, that is, if we denote the points by P_1 and P_2, the distance is given by

$$\int_{P_1}^{P_2} (dx^2 + dy^2 + K^{-1} dz^2)^{1/2}, \tag{1.26}$$

where the path of integration lies in the intersection of S_K with the plane OP_1P_2. Via central projection each of these metrics induces one on the plane $z = 1$. Since the sphere of radius $K^{-1/2}$ gets flatter as K approaches zero, it is geometrically clear that this induced metric on the plane $z = 1$ approaches the Euclidean one. As before, we have the formula

$$\text{distance}(P_1, P_2) = \frac{1}{2iK^{1/2}} \log[(x_1, y_1, K^{-1/2}z_1), (x_2, y_2, K^{-1/2}z_2), R_1, R_2],$$

where R_1 and R_2 are the points at which the line through $(x_1, y_1, K^{-1/2}z_1)$ and $(x_2, y_2, K^{-1/2}z_2)$ meets the absolute

$$x^2 + y^2 + z^2 = 0.$$

Now the roots of

$$(x_1 + \lambda x_2)^2 + (y_1 + \lambda y_2)^2 + (K^{-1/2}z_1 + \lambda K^{-1/2}z_2)^2 = 0$$

are simply those λ such that

$$P_1 + \lambda P_2$$

lies on the complex cone

$$K(x^2 + y^2) + z^2 = 0.$$

Thus for the metric induced on the plane $z = 1$ by central projection, we have

$$\text{distance}(p_1, p_2) = \frac{1}{2iK^{1/2}} \log(p_1, p_2, r_1, r_2), \tag{1.27}$$

where r_1, r_2 form the intersection of the line through p_1, p_2 with the conic

$$K(x^2 + y^2) + 1 = 0.$$

Since the metric (1.27) is induced from the standard metric on a sphere of radius $K^{-1/2}$, it is a metric of constant curvature K.

1.8 Through the Looking Glass

We have seen that the metric (1.26) *when restricted to S_K* stays bounded as K approaches 0, so we can in fact let K become negative and the formula (1.26) will still give a perfectly good metric on S_K (or at least on the "top sheet" of what is now a hyperboloid). By "analytic continuation" we still expect that this will be a metric of constant curvature K and that geodesics are intersections of S_K with planes through 0. Indeed this is the case. By central projection this induces a metric on the part of the $z = 1$ plane which is interior to the circle

$$K(x^2 + y^2) + 1 = 0.$$

Again by analytic continuation the metric is still given by the formula (1.27), where now all four points p_1, p_2, r_1, r_2 are in the *real* plane $z = 1$. In fact, we have Figure 1.13. Now any linear transformation of $\mathbb{R}^3$ which takes

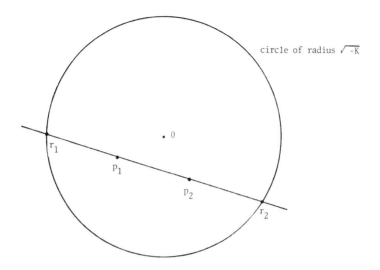

circle of radius $\sqrt{-K}$

Figure 1.13. Hyperbolic geometry in the disk of radius $(-K)^{1/2}$.

S_K onto itself induces an isometry since, perforce, it preserves the real symmetric bilinear form

$$K(x^2 + y^2) + z^2. \tag{1.28}$$

To study the group of transformations preserving (1.28) it suffices to consider the cases

$$K = 1, \qquad K = -1,$$

since the only important thing about the symmetric real bilinear form (1.28) is its number of positive and negative eigenvalues. We have already considered the case $K = 1$. In the case $K = -1$ the group of isometries is called the *Lorentz group* because of its importance in the theory of relativity. To see how it acts, notice that it contains all rotations around the z axis and also that it contains transformations which leave y fixed and take

$$(1, y, 0) \mapsto (x_1, y, z_1),$$
$$(0, y, 1) \mapsto (x_2, y, z_2)$$

such that

$$z_1^2 - x_1^2 = -1,$$
$$z_2^2 - x_2^2 = 1,$$
$$z_1 z_2 - x_1 x_2 = 0.$$

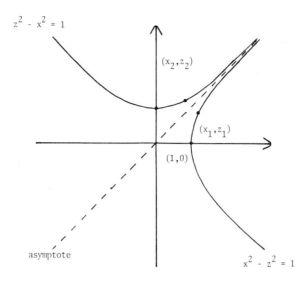

Figure 1.14. A Lorentz transformation moving along a system of hyperbolas.

In the (x, z) plane, we have Figure 1.14. Since the involution which interchanges the x axis and the z axis must interchange the points (x_1, z_1) and (x_2, z_2), they must be "equidistant" along their respective hyperbolas. From this it is immediate that the group of Lorentz transformations is transitive on the hyperboloid S_K and also transitive on the set of directions through any fixed point on S_K. Thus our metric must indeed be of constant (negative) curvature.

1.9 The Polar Curve

There's another very useful general concept which can be applied in this setting, namely the concept of the *polar curve* associated to a given plane curve. Namely, if a plane curve C in $\mathbb{CP}_2$ is given by the homogeneous equation

$$F(x, y, z) = \sum_{i=0}^{m} r_i(x, y)z^i = 0,$$

then the condition that the line

$$z = 0$$

be tangent to C at $(1, 0, 0)$ is simply that the equation

$$r_0(x, y) = 0$$

have a double root at $(1, 0)$, that is, that

$$r_0(1, 0) = 0, \qquad \frac{\partial r_0}{\partial y}(1, 0) = 0.$$

This condition can be restated as

$$\frac{\partial r_0}{\partial x}(1, 0) = 0, \qquad \frac{\partial r_0}{\partial y}(1, 0) = 0$$

by *Euler's formula*, which says that

$$mF = x\,\frac{\partial F}{\partial x} + y\,\frac{\partial F}{\partial y} + z\,\frac{\partial F}{\partial x},$$

$$(\deg r)r = x\,\frac{\partial r}{\partial x} + y\,\frac{\partial r}{\partial y}.$$

We can rewrite the tangency condition simply as

$$\frac{\partial F}{\partial x}(1, 0, 0) = \frac{\partial F}{\partial y}(1, 0, 0) = 0.$$

Since any point in $\mathbb{C}\mathbb{P}_1$ can be moved to $(1, 0, 0)$ by a linear transformation in such a way that a given line through the point goes to the line $z = 0$, we see in general that a line

$$ax + by + cz = 0$$

is tangent to the curve

$$F(x, y, z) = 0$$

at (x_0, y_0, z_0) if and only if $(x_0, y_0, z_0) \in C$ and

$$\left. \left(\frac{\partial F}{\partial x}, \frac{\partial F}{\partial y}, \frac{\partial F}{\partial z} \right) \right|_{(x_0,\, y_0,\, z_0)} = \lambda(a, b, c) \qquad (1.29)$$

for some $\lambda \in \mathbb{C}$. [If the left-hand triple vanishes, we will say that every line through (x_0, y_0, z_0) is tangent to C at that point. If the left-hand side does not vanish, the fact that $ax_0 + by_0 + cz_0 = 0$ is insured by Euler's formula.]

The mapping

$$\mathcal{D}_C: \qquad \mathbb{C}\mathbb{P}_2 \longrightarrow \mathbb{C}\mathbb{P}_2,$$

$$(x_0, y_0, z_0) \longrightarrow \left. \left(\frac{\partial F}{\partial x}, \frac{\partial F}{\partial y}, \frac{\partial F}{\partial z} \right) \right|_{(x_0,\, y_0,\, z_0)} \qquad (1.30)$$

is called the *polar mapping* associated with the plane curve C. $\mathscr{D}_C(C)$ is called the *dual curve* of C, which we denote $\hat{C}$. If the curve C is a nondegenerate conic, $\mathscr{D}_C$ is of course a linear isomorphism so $\hat{C}$ is again a nondegenerate conic. To obtain its equation, notice that if C is given by the matrix equation

$$vA\,{}^t v = 0,$$

then $\hat{C}$ is given by a matrix M such that

$$AMA = A.$$

(Remember that A is symmetric.) So

$$M = A^{-1}$$

and

$$\hat{\hat{C}} = C. \tag{1.31}$$

Now let C be a nondegenerate plane curve of any degree. (Nondegenerate means that the three partials of the homogeneous defining equation F do not vanish simultaneously.) If $\mathscr{D}_C$ is locally of maximal rank, the equation $G = 0$ of the dual curve $\hat{C}$ is determined locally by the formula

$$G\left(\frac{\partial F}{\partial x},\ \frac{\partial F}{\partial y},\ \frac{\partial F}{\partial z}\right) = \lambda F(x,\, y,\, z).$$

Differentiating, we get that if the matrix of second partials

$$\begin{bmatrix} \partial^2 F/\partial x^2 & \partial^2 F/\partial x\,\partial y & \partial^2 F/\partial x\,\partial z \\ \partial^2 F/\partial x\,\partial y & \partial^2 F/\partial y^2 & \partial^2 F/\partial y\,\partial z \\ \partial^2 F/\partial x\,\partial z & \partial^2 F/\partial y\,\partial z & \partial^2 F/\partial z^2 \end{bmatrix} \tag{1.32}$$

is of maximal rank at $(x_0,\, y_0,\, z_0) \in C$, then the point

$$\mathscr{D}_C\left(\frac{\partial F}{\partial x},\ \frac{\partial F}{\partial y},\ \frac{\partial F}{\partial z}\right)$$

depends only on the values of the first and second partials of F at (x_0, y_0, z_0). Since there exists a conic whose defining equation has the same value (namely, zero) and the same first and second partials at (x_0, y_0, z_0) that F does, the formula (1.31) for conics implies the same formula for general plane curves. To be more precise, if the mapping $\mathscr{D}_C$ is of maximal rank at a point $P \in \mathbb{CP}_2$, then in a neighborhood of P on C

$$\mathscr{D}_{\hat{C}} \circ \mathscr{D}_C = \text{identity map.}$$

Via the linearity of the polar mapping in the case of conics, every fact about conics has a corresponding "dual fact." To begin, let A be the symmetric matrix associated with the nondegenerate conic C. The simultaneous equations

$$[x \quad y \quad z]A \begin{bmatrix} x \\ y \\ z \end{bmatrix} = 0, \tag{1.33}$$

$$[x \quad y \quad z]A \begin{bmatrix} x_0 \\ y_0 \\ z_0 \end{bmatrix} = 0, \tag{1.34}$$

for fixed $(x_0, y_0, z_0) \in \mathbb{CP}_2$, must have two solutions:

$$(x_1, y_1, z_1) \quad \text{and} \quad (x_2, y_2, z_2).$$

Since we have seen that the equation of the tangent line to C at (x_i, y_i, z_i) is

$$[x_i \quad y_i \quad z_i]A \begin{bmatrix} x \\ y \\ z \end{bmatrix} = 0,$$

the point (x_0, y_0, z_0) must constitute the intersection of the tangent lines to C at (x_1, y_1, z_1) and (x_2, y_2, z_2) (Figure 1.15). Thus the polar mapping $\mathcal{D}_C$ can be thought of as assigning to $(x_0, y_0, z_0) \in \mathbb{CP}_2$ the line (1.34) which meets C in the two points whose tangents pass through (x_0, y_0, z_0). From this it is clear that we have a nice bijection between $\mathbb{CP}_2$ and

$$C^{(2)},$$

the set of pairs of points on C (often called the second symmetric product

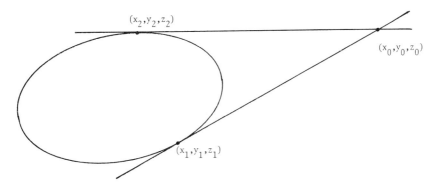

Figure 1.15. Constructing the polar of a point.

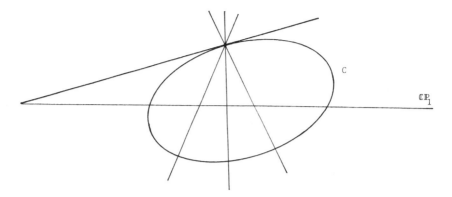

Figure 1.16. Stereographic projection of a conic onto a projective line.

of C with itself). Since C is bijective to $\mathbb{CP}_1$ via stereographic projection (Figure 1.16), we see that there is a nice bijection

$$\mathbb{CP}_2 \cong (\mathbb{CP}_1)^{(2)}.$$

All this is consistent with the fact we saw earlier: if $(x_0, y_0, z_0) \in C$, then its polar line is tangent to C at (x_0, y_0, z_0).

Our polar isomorphism then sends points to lines (and lines to points) in a way which is full of geometric meaning. The "polar" or "dual" statements of

> point p on C
> two lines passing through a point

are

> line tangent to C at p
> two points spanning a line.

Thus, for example, Pascal's theorem on the "mystic hexagon" which we saw earlier translates to the dual theorem:

> *The lines connecting opposite vertices of a hexagon circumscribed about a conic pass through a common point.*

1.10 Perpendiculars in Hyperbolic Space

Let's see an application of the polar mapping to the plane geometries of constant curvature that we studied earlier—the so-called *spherical* and *hyperbolic* geometries. These consisted of sets of points with a distance function. The distance function was very special in the sense that each geometry has the largest possible group of isometries, namely, given any point and direction at that point there is a (unique) isometry which takes them to any other point and direction. By central projection we achieved a planar representation of each geometry (or, in the case $K > 0$, at least a planar representation of a big piece of it). One difficulty with this representation is the notion of angle. For example, in the case $K < 0$ the usual plane Euclidean notion of angle between two lines in our geometry is no good because, for example, the law of the cosines does not even hold infinitesimally and isometries do not preserve angles. We need a way to see geometrically what angles are in this geometry—or, failing that, at least we ought to be able to decide geometrically when two lines are perpendicular. We begin with a sphere of radius $K^{-1/2}$ centered about the origin in $\mathbb{R}^3$. Consider two geodesics on the sphere through a point $P = (x_0, y_0, z_0)$ (Figure 1.17). Let M and M' be the intersections of these two geodesics with the geodesic

$$x_0 x + y_0 y + z_0 z = 0 \qquad (1.35)$$

on the sphere. If δ is the angle between the two original geodesics, then we have seen that

$$K^{-1/2}\delta = \frac{1}{2iK^{1/2}} \log(M, M', N, N'),$$

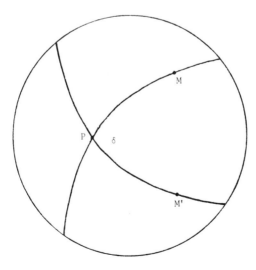

Figure 1.17. The angle between geodesics in spherical geometry.

where N, N' are the points of intersection of the complex line (1.35) with the absolute

$$x^2 + y^2 + z^2 = 0, \tag{1.36}$$

that is, N and N' are the two points of the conic (1.36) whose tangents pass through P! If we translate all this as we did before to a statement about the K-geometry in the plane

$$z = 1,$$

we obtain that the K-angle between the lines L and L' through $p = (x, y)$ is given by

$$\frac{1}{2i} \log(L, L', J, J'),$$

where J and J' are the two tangent lines to

$$K(x^2 + y^2) + 1 = 0$$

which pass through p. (The cross ratio of four lines passing through a point p is simply the cross ratio of their four points of intersection with another line not containing p.) By analytic continuation we expect the same formula to hold for negative K, which it does. We cannot "see" the lines J and J' in either the case of K positive or K negative (i.e., the lines J and J' are not real). However, we can see what it means for two geodesics to be perpendicular.

Let's get at perpendicularity, then, in the case $K < 0$. We have two lines L and L' (geodesics) intersecting in the point p_0 lying in the interior of the disk bounded by

$$K(x^2 + y^2) + 1 = 0 \tag{1.37}$$

(Figure 1.18). We have seen that each line has an associated polar point, call them p and p', with respect to the conic (1.37)—p is simply the point at which the tangents to (1.37) at its intersection with L meet. L and L' will be perpendicular if $\log(L, L', J, J') = \pi i$, that is, if

$$(L, L', J, J') = -1.$$

But, if A is the matrix of a conic as in Section 1.9,

$$pA\,{}^t p_0 = 0$$

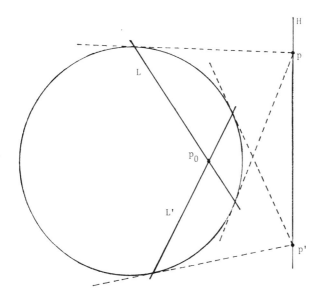

Figure 1.18. Angles in hyperbolic geometry.

means that $p \in$ (polar line of p_0) which we will call H. Also $p' \in H$. Under the polar mapping we have

$$\text{point } p \to \text{line } L,$$
$$\text{point } p' \to \text{line } L',$$
$$\text{point } p_0 \to \text{line } H,$$
$$\text{point } n \to \text{line } J,$$
$$\text{point } n' \to \text{line } J',$$

where n and n' are the points at which J and J' intersect the conic (1.37). But n and n' must lie on H since J and J' pass through p_0. The polar mapping is linear, so it preserves cross ratio. Thus

$$(L, L', J, J') = (p, p', n, n').$$

Suppose now that $p \in L'$. Then

$$pA{}^tp' = 0$$

so that $p' \in L$. In this case

$$(L, L', J, J') = (p', p, n, n'),$$

since we can compute the cross ratio of four lines through p_0 by computing that of their intersection with the line H. Thus if $p \in L$ (or equivalently $p' \in L$), then

$$(p', p, n, n') = (p, p', n, n').$$

Since p, p', n, n' are all distinct, the only way that this is possible is if

$$(p, p', n, n') = -1.$$

Conversely, since (L, L', J, J') is an injective function of L', it is clear that if $(L, L', J, J') = -1$, then $p \in L'$ (and so also $p' \in L$). Thus

L and L' are anharmonic with respect to J and J' [i.e., $(L, L', J, J') = -1$] if and only if L and L' meet perpendicularly in the K-geometry.

In totally geometric terms:

The perpendicular to L passing through p_0 is the line $\overline{p_0\,p}$, where p is the polar pont of L.

1.11 Circles in the K-Geometry

PROBLEM: Find the set of points of fixed distance from a given point p_0.

This problem is solved by finding a curve which goes into itself under the isometries which fix p_0. We let L be the polar line of p_0, which intersects the conic (1.37) in the (imaginary) points n and n'. If

$$J(x, y, z) = 0, \qquad J'(x, y, z) = 0$$

are the equations of the two tangent lines to (1.37) at n and n' respectively, then we have the family of conics

$$t_0[K(x^2 + y^2) + 1] + t_1 J(x, y, z)J'(x, y, z) = 0$$

parametrized by the homogeneous coordinates (t_0, t_1). We will denote this family by

$$C_{(t_0, t_1)}, \qquad (t_0, t_1) \in \mathbb{CP}_1. \tag{1.38}$$

Just as there is, in general, only a $\mathbb{CP}_1$ of conics passing through four points, there is only a $\mathbb{CP}_1$ of conics passing through two points and

having fixed tangent direction at each. So the family (1.38) must contain
the degenerate conic consisting of the line

$$L = nn'$$

counted twice. Now the isometries of our geometry are the linear transformations which leave the conic (1.37) fixed. In particular, we have a one-
real-parameter family of isometries which leave p_0 (and therefore n and n')
fixed and which act transitively on the set of real lines through p_0. This
group of isometries acts linearly on the family of conics (1.38) but in fact
leaves three conics [$2 \cdot L$, the conic (1.37), $\overline{(p_0 \, n)} + \overline{(p_0 \, n')}$] fixed. Therefore
this group acts trivially on the set of conics (1.38). From these facts it
follows immediately that the family

$$t_0[K(x^2 + y^2) + 1] + t_1 \left[\begin{bmatrix} x & y & 1 \end{bmatrix} \begin{bmatrix} K & 0 & 0 \\ 0 & K & 0 \\ 0 & 0 & -1 \end{bmatrix} {}^t p_0 \right]^2 = 0, \quad (1.39)$$

$t_0, t_1 \in \mathbb{R}$, cuts out a family of "circles" in K-geometry, that is, each conic
in (1.39) which hits the set

$$(x, y) \text{ real}, \qquad K(x^2 + y^2) + 1 > 0,$$

meets it in a set of points of fixed distance from p_0. The "dual" formula-
tion of the problem, starting with a line L in K-geometry and its polar
point p_0 outside the disk of points of the geometry, gives rise to the conclusion that the family (1.39) consists of loci of points of fixed distance from
the line L (Figure 1.19).

In any geometry, a curve has assigned to it at each of its points a
number, called its *geodesic curvature* at that point, which depends only on
the distance function of the geometry (O'Neill [7], pp. 329–330). In a
geometry with as many isometries as the K-geometry, $K < 0$, circles therefore must have the same geodesic curve at each of their points. To see how
this curvature behaves, let's consider Figure 1.20. Consider the "K-circle"
of center $(a, 0)$ which passes through $(0, 0)$. As a approaches $(-K)^{-1/2}$, this
circle approaches a conic whose four points of contact with
$K(x^2 + y^2) + 1 = 0$ all coalesce at $((-K)^{-1/2}, 0)$. [To see this, examine our
geometric construction of the family (1.38) or the Figure 1.20.] We say that
the limiting conic and $K(x^2 + y^2) + 1 = 0$ have contact of order 4 at
$((-K)^{-1/2}, 0)$. The order of contact is so great (≥ 3) that the usual
Euclidean curvature of these two curves must be the same at this point.
The common value is therefore

$$(-K)^{1/2}.$$

By symmetry the limiting conic has Euclidean curvature $(-K)^{1/2}$ at $(0, 0)$.

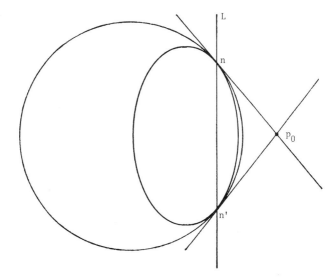

Figure 1.19. The locus of points of fixed distance from a line in hyperbolic geometry.

But now our K-metric and the usual Euclidean metric

$$\int (dx^2 + dy^2)^{1/2}$$

can be shown to coincide to second order at $(0, 0)$. In differential geometry it is shown that this implies that the geodesic curvatures with respect to the

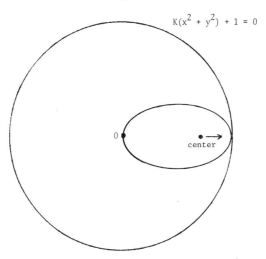

Figure 1.20. Limit of circles through a fixed point in hyperbolic geometry.

two metrics must coincide at (0, 0). We can therefore conclude that in K-geometry there are no circles of geodesic curvature less than $(-K)^{1/2}$ but that all curvature values greater than $(-K)^{1/2}$ are attained!

It's now time to go on to other things, but before we wind up our treatment of conics, there is one more topic we ought to touch on.[†]

1.12 Rational Points on Conics

We should say a few words about the number theory of conics. That is, suppose we are given an equation

$$Ax^2 + Bxy + Cy^2 + Dxz + Eyz + Fz^2 = 0 \qquad (1.40)$$

where the coefficients $A, B, \ldots, F \in \mathbb{Q}$, the field of rational numbers. We ask when the equation (1.40) has a solution in

$$\mathbb{Q}\mathbb{P}_2,$$

the set of one-dimensional subspaces of the vector space $\mathbb{Q}_3$. If there is one such solution

$$(x_0, y_0, z_0),$$

then we can proceed by stereographic projection (Figure 1.21) to construct many others as long as the set of complex solutions, that is, solutions in $\mathbb{C}\mathbb{P}_2$, is a nice, nondegenerate conic. For suppose $y_0 \neq 0$; then given any point $(x_1, 0, z_1) \in \mathbb{Q}\mathbb{P}_2$ on the line $y = 0$, the equation

$$A(x_0 + tx_1)^2 + B(x_0 + tx_1)y_0 + Cy_0^2 + D(x_0 + tx_1)(z_0 + tz_1)$$
$$+ Ey_0(z_0 + tz_1) + F(z_0 + tz_1)^2 = 0 \quad (1.41)$$

has the solution $t = 0$ and therefore has a second rational solution. So the problem of deciding what the solution set of (1.40) in $\mathbb{Q}\mathbb{P}_2$ is like really rests with answering the single question: Does (1.40) have *any* rational solutions? Clearing denominators, we can assume that $A, B, C, \ldots, F$ are integers and ask the equivalent question: Does (1.40) have any *integral* solutions?

The beautiful thing is that this last question can be effectively decided for any given equation (1.40). First, the process we described earlier of finding

[†] For the reader who wants to delve deeper into Riemannian geometry, there is perhaps no more beautiful source than the classic *Leçons sur la Géométrie des Espaces de Riemann* of Élie Cartan (Paris: Gauthier-Villars, 1963).

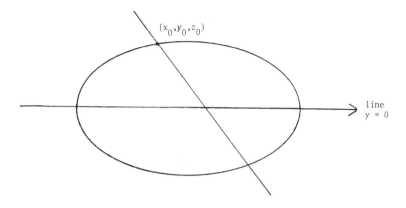

Figure 1.21. Stereographic projection again.

a linear transformation of $\mathbb{RP}_2$ which transforms our equation (1.40) to one of the form

$$\varepsilon_1 x^2 + \varepsilon_2 y^2 - z^2 = 0 \qquad (1.42)$$

can actually be accomplished over $\mathbb{Q}$, that is, by a 3×3 matrix with rational entries which induces an automorphism of $\mathbb{QP}_2$, as long as we do not require that $|\varepsilon_i| = 1$. So we can assume that our equation has the form (1.42). Since to have rational solutions to (1.42) we must have real ones, one of the ε_i must be positive. Also we can assume that $|\varepsilon_1| \geq |\varepsilon_2|$. Replacing x by ax and y by by, we can further assume that $|\varepsilon_i|$ is a product of distinct primes each raised to the first power.

Let's look at two examples:

$$3x^2 + 2y^2 - z^2 = 0, \qquad (1.43)$$

$$3x^2 + y^2 - z^2 = 0. \qquad (1.44)$$

In (1.43), suppose we had a solution (x_0, y_0, z_0) of integers with no common factor. Then if $3 \nmid y_0$ (3 does not divide y_0), we obtain a contradiction by regarding (1.43) as an equation with coefficients in the field

$$\mathbb{F}_3 = \mathbb{Z}/3\mathbb{Z}$$

of integers modulo 3. But if $3 \mid y_0$, then $3 \mid z_0$ so $3^2 \mid 3x_0^2$ so $3 \mid x_0$. Since x_0, y_0, and z_0 have no common factor, (1.43) has no rational solutions. On the other hand, the equation (1.44) admits the solution

$$(1, 1, 2).$$

Notice that in (1.44) ε_2 is a square modulo ε_1.

This sort of reasoning leads to the conclusion, in general, that if (1.42) has an integral solution, then

$$\varepsilon_2 \text{ is a square modulo } p$$

for each prime number p dividing ε_1. So by the Chinese remainder theorem

$$\varepsilon_2 \text{ is a square modulo } \varepsilon_1.$$

This means that there exist integers c, ε'_1 such that $|c| \leq |\varepsilon_1|/2$ and

$$c^2 = \varepsilon_2 + \varepsilon_1 \varepsilon'_1,$$
$$\varepsilon_1 \varepsilon'_1 = c^2 - \varepsilon_2 = (c + \varepsilon_2^{1/2})(c - \varepsilon_2^{1/2}).$$

So, by algebra, the existence of rational numbers

$$a, b$$

such that

$$\varepsilon_1 = (a + b\varepsilon_2^{1/2})(a - b\varepsilon_2^{1/2})$$

is equivalent to the existence of rational numbers

$$a', b'$$

such that

$$\varepsilon'_1 = (a' + b'\varepsilon_2^{1/2})(a' - b'\varepsilon_2^{1/2}).$$

In other words,

$$\varepsilon_1 x^2 + \varepsilon_2 y^2 - z^2 = 0$$

has a rational solution if and only if

$$\varepsilon'_1 x^2 + \varepsilon_2 y^2 - z^2 = 0$$

does. If $|\varepsilon_1| > 1$, then

$$|\varepsilon'_1| = \left| \frac{c^2 - \varepsilon_2}{\varepsilon_1} \right| \leq \frac{|\varepsilon_1|}{4} + 1 < |\varepsilon_1|,$$

so that (eliminating squares which appear in the factorization of $|\varepsilon'_1|$) we have reduced the problem to one in which $|\varepsilon_1| + |\varepsilon_2|$ is smaller. We can repeat this argument until $|\varepsilon_2|$ fails to be a square modulo $|\varepsilon_1|$ or until we reach the case

$$|\varepsilon_1| = |\varepsilon_2| = 1,$$

which admits a solution if and only if ε_1 or ε_2 is positive.

Cubics

2.1 Inflection Points

Let us now take up the study of the solution set $E \subseteq \mathbb{CP}_2$ of the equation

$$F(x, y, z) = 0, \tag{2.1}$$

where F is a homogeneous polynomial of degree 3. Again we will assume that the partial derivatives

$$\frac{\partial F}{\partial x}, \frac{\partial F}{\partial y}, \frac{\partial F}{\partial z} \tag{2.2}$$

do not all vanish simultaneously. Unlike the degree-2 case, every cubic polynomial with real coefficients has at least one real root, so if the coefficients of F are real, then the solution set to (2.1) in $\mathbb{RP}_2$ will always be a smooth curve. The question whether there exists a solution in $\mathbb{QP}_2$ if the coefficients of (2.1) are rational is extremely difficult, and there is as yet no known procedure for deciding in general.

First, let's look at the polar mapping

$$\mathscr{D}_E \colon \mathbb{CP}_2 \to \mathbb{CP}_2$$

which we introduced in Chapter One. When the degree of E is three, this mapping is no longer an isomorphism. In fact, it is degenerate at points p where

$$\det \begin{bmatrix} \dfrac{\partial^2 F}{\partial x^2}(p) & \dfrac{\partial^2 F}{\partial x\,\partial y}(p) & \dfrac{\partial^2 F}{\partial y\,\partial z}(p) \\[3ex] \dfrac{\partial^2 F}{\partial x\,\partial y}(p) & \dfrac{\partial^2 F}{\partial y^2}(p) & \dfrac{\partial^2 F}{\partial y\,\partial z}(p) \\[3ex] \dfrac{\partial^2 F}{\partial x\,\partial z}(p) & \dfrac{\partial^2 F}{\partial y\,\partial z}(p) & \dfrac{\partial^2 F}{\partial z^2}(p) \end{bmatrix} = 0. \tag{2.3}$$

Since this determinantal equation is homogeneous of degree 3, we expect that there will be $3 \times 3 = 9$ points of E where the mapping is, in fact, degenerate. But what about

$$\mathcal{D}_E|_E?$$

Suppose we are at a point $p \in E$ at which (2.3) holds. If $\mathcal{D}_E|_E$ were to be of maximal rank at p, then the kernel of $\mathcal{D}_E$ at p could not lie in the tangent space to E at p. This would imply the existence of $[x \quad y \quad z]$ such that

$$[x \quad y \quad z] \cdot {}^t\!\left(\frac{\partial F}{\partial x}(p), \quad \frac{\partial F}{\partial y}(p), \quad \frac{\partial F}{\partial z}(p) \right) \neq 0$$

but such that

$$[x \quad y \quad z] \cdot M = (0, 0, 0),$$

where M is the matrix in (2.3). But multiplying this last equation on the right by ${}^t\!p$, we obtain a contradiction by Euler's formula. Since the mapping $\mathcal{D}_E|_E$ is clearly not everywhere degenerate, we see that it is degenerate at most nine points. Now suppose we have changed coordinates so that one of these points is

$$p_0 = (0, 1, 0)$$

and at p_0,

$$\left(\frac{\partial F}{\partial x}, \frac{\partial F}{\partial y}, \frac{\partial F}{\partial z} \right) = (0, 0, 1).$$

Then the equation of E takes the form

$$y^2 z + yQ(x, z) + C(x, z) = 0, \tag{2.4}$$

where Q and C are homogeneous forms of degree 2 and degree 3 respectively. Let

$$Q(x, z) = \tfrac{1}{2}(ax^2 + 2bxz + cz^2).$$

The degeneracy of $\mathcal{D}_E$ at p_0 is simply the degeneracy of the matrix

$$\begin{bmatrix} ay + \dfrac{\partial^2 C}{\partial x^2} & ax + bz & by + \dfrac{\partial^2 C}{\partial x\, \partial z} \\[2ex] ax + bz & 2z & 2y + bx + cz \\[2ex] by + \dfrac{\partial^2 C}{\partial x\, \partial z} & 2y + bx + cz & cy + \dfrac{\partial^2 C}{\partial z^2} \end{bmatrix} \tag{2.5}$$

at $(0, 1, 0)$. Thus p_0 is an inflection point if and only if

$$a = 0.$$

Next we differentiate

$$\left.\frac{\partial[\text{determinant of } (2.5)]}{\partial x}\right|_{p_0} = -2\frac{\partial^3 C}{\partial x^3}.$$

But

$$\frac{\partial^3 C}{\partial x^3} \neq 0,$$

since otherwise the equation (2.4) would have no x^3 term and the curve would contain the line $z = 0$ and so would be singular. Thus the zero set of the determinant of (2.5) and E meet transversely at p_0. From this we can conclude that $\mathcal{D}_E$ is degenerate at nine *distinct* points of E. We also see that if we restrict the equation (2.4) to the line

$$z = 0,$$

then it becomes

$$ex^3 = 0,$$

so we say that the line and E *have contact of order* 3 at p_0. This last fact is equivalent to the degeneracy of $\mathcal{D}_E$ at p_0, an equivalency which is easily seen to continue to hold for equations and curves of degree higher than three. The same is not true of the computation which allowed us to conclude that the Hessian curve defined by the determinant of

$$\left[\frac{\partial^2 F}{\partial x_i \partial x_j}\right]$$

meets E simply (or transversely) at all its points of intersection. These points of intersection are called the *inflection points* of the curve, and there are

$$n[3(n-2)]$$

of them, where $n = $ degree of the curve.

2.2 Normal Form for a Cubic

Now suppose we have a nondegenerate (or, as we shall sometimes call it, a nonsingular) curve E of degrees 3, and suppose we have arranged things so that one of the nine inflection points of E is the point

$$p_\infty = (0, 1, 0)$$

and the tangent line to E at that point is the line

$$z = 0,$$

that is, the line "at infinity" with respect to the affine-coordinates (x, y).

Let us look now at the family of lines

$$x = ez, \qquad e \in \mathbb{C}. \qquad (2.6)$$

By our assumption about $p_\infty \in E$, the equation of E must have the form

$$y^2z + yQ(x, z) + C(x, z),$$

and we can compute those values of e for which the line (2.6) is tangent to E by computing the discriminant of the equation

$$y^2 + Q(e, 1)y + C(e, 1) = 0,$$

that is, the determinant of

$$\begin{bmatrix} 1 & Q(e, 1) & C(e, 1) \\ 2 & Q(e, 1) & 0 \\ 0 & 2 & Q(e, 1) \end{bmatrix};$$

or, simply, we find the set of solutions to

$$-Q(e, 1)^2 + 4C(e, 1) = 0. \qquad (2.7)$$

Since we have seen that $\partial^3 C/\partial x^3 \neq 0$ and $\partial^2 Q/\partial x^2 = 0$, this equation has three roots (counting, for the moment, possible multiple roots). Assume that one of these roots corresponds to the point

$$(0, 0, 1).$$

Then the line $x = 0$ is tangent to E at that point so that $\partial^3 C/\partial z^3 = \partial^2 Q/\partial z^2 = 0$. If e is to be a multiple root of (2.7), then we must further have that

$$\frac{\partial^3 C}{\partial x\, \partial z^2} = 0.$$

But then all the partial derivatives of the defining equation of E vanish at $(0, 0, 1)$, contradicting the nonsingularity of E. Thus the roots of (2.7) are distinct and correspond to three distinct points

$$p_1, p_2, p_3$$

in $\mathbb{CP}_2$. The picture is shown in Figure 2.1.

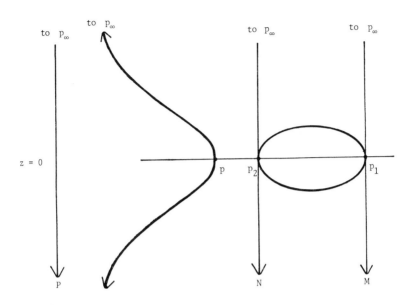

Figure 2.1. Projecting a cubic from the inflection point p_∞.

Now let us consider three cubic curves:

$$E,$$
$$(\text{line } M) + (\text{line } N) + (\text{line } p_\infty p),$$
$$(\text{line } L) + 2(\text{line } p_1 p_2),$$

where p is the third point of intersection of E with the line through p_1 and p_2. The three cubic curves in the list have eight common points, namely,

$$3p_\infty, \, 2p_1, \, 2p_2, \, p.$$

But, as we saw in Chapter One when we were trying to cope with Pascal's mystic hexagon, we therefore expect the defining equations of the three cubics to be linearly dependent, which means that the three cubics have a ninth point in common. This point must be p, so that the line $(p_\infty p)$ must meet E twice at p, that is, $p = p_3$ and the three points p_1, p_2, p_3 lie on a line.

By appropriate linear change of coordinates, which disturbs nothing we have done so far, we can assume

$$p_1 = (0, 0, 1), \qquad p_2 = (1, 0, 1), \qquad p_3 = (\lambda, 0, 1).$$

Since then $Q(e, 1)$ must vanish at

$$e = 0, 1, \lambda,$$

we conclude that $Q(x, z) \equiv 0$ and so our equation of E comes to have the form

$$y^2 z - x(x - z)(x - \lambda z) = 0. \tag{2.8}$$

2.3 Cubics as Topological Groups

When we were studying conics in Chapter One we scarcely mentioned their topology in $\mathbb{CP}_2$. This was because everything was so easy. Namely, via stereographic projection (Figure 2.2) we set up a diffeomorphism between our conic and a projective line L, which is in turn diffeomorphic to the standard two-sphere in $\mathbb{R}^3$. When we attempt stereographic projection with an elliptic curve, however, things become a bit more complicated (see Figure 2.1.) If the center of projection p_∞ is an inflection point, then the equation (2.7) is of degree 3; but, if not, it will be of degree 4. Stereographic

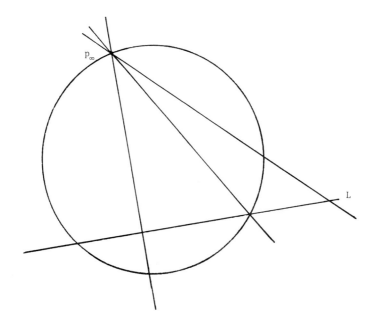

Figure 2.2. Projecting a conic from a point p_∞ on it.

projection then presents E as a two-sheeted covering of $\mathbb{CP}_1$ "ramified" at four points (one of which is at infinity if p_∞ is an inflection point). What "ramified" means is that around the point in question the projection map has the same behavior as does the projection

$$(\text{curve } y^2 = x) \longrightarrow (x \text{ axis}),$$

$$(x, y) \longrightarrow x$$

near $(0, 0)$. From this it is easy to make a topological model for E. Take two spheres, slit each twice in identical ways (Figure 2.3), open up the slits to make holes, and then turn the bottom sphere over and paste it to the top along the edges of the slits so that the markings match (Figure 2.4). Thus E is topologically a torus, that is, the quotient space

$$\mathbb{R} \times \mathbb{R} / \mathbb{Z} \times \mathbb{Z}.$$

So E is a topological group, in fact.

But of course the group structure on E is much more intimately related with its complex geometry. Let's now explore this relationship. We can get a good idea of what is going on by assuming that λ is real and by looking at the solution set to (2.8) in $\mathbb{RP}_2$ or, simply, in the (x, y)-plane $\mathbb{R}^2$ (see Figure 2.5). We define a binary operation on E by defining

$$p_1 + p_2 = (x, -y),$$

where p_1, p_2, and (x, y) lie on a line. There is no problem that this binary operation is commutative and that each element has an inverse if we define the point at infinity, p_∞, as the identity element. The interesting verification

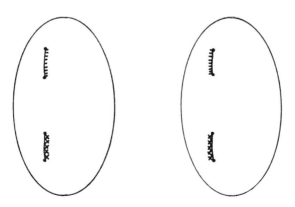

Figure 2.3. Cutting a cubic into two pieces.

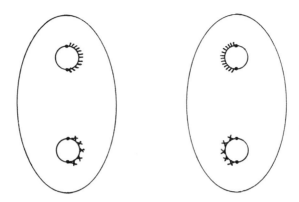

Figure 2.4. Pasting the cubic back together again.

is that of the associativity of the operation. The main idea—that cubics passing through eight points have a ninth in common—is one that we have already used twice before. To see how it works in this case, look at Figure 2.6, a diagram borrowed from John Tate's beautiful lectures at Haverford College in April 1961 on the number theory of cubic curves. To show associativity we must show that the points $(p + q)r$ and $p(q + r)$ coincide.

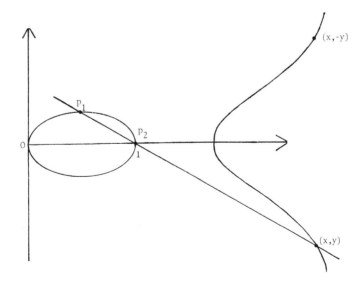

Figure 2.5. A real cubic in normal form, cut by a line.

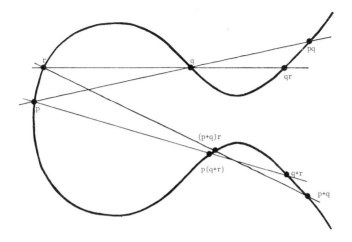

Figure 2.6. The associativity of addition on a cubic.

So we must find two cubic curves (besides E) that pass through the eight points

$$p_\infty, \; p, \; q, \; r, \; qr, \; pq, \; (q + r), \; (p + q).$$

Then if one of these cubics passes through $(p + q)r$ and the other through $p(q + r)$, we can conclude that the two coincide. The two cubic curves in question are

> (line through p and q)
> + (line through qr and $(q + r)$)
> + (line through r and $(p + q)$)

and

> (line through q and r)
> + (line through pq and $(p + q)$)
> + (line through p and $(q + r)$).

2.4 The Group of Rational Points on a Cubic

So it should be clear that the complex solution set to a nondegenerate cubic

$$F(x, y, z) = 0 \tag{2.9}$$

forms a group with a geometrically constructed group operation with any given inflection point as identity element. Now suppose the coefficients in (2.9) are real or rational and that we have a solution $p \in \mathbb{RP}_2$ or $p \in \mathbb{QP}_2$. Now p may or may not be an inflection point, and no *linear* change of coordinates will change p into an inflection point if it is not already one. However, there is a birational transformation of $\mathbb{RP}_2$ or $\mathbb{QP}_2$ (and so $\mathbb{CP}_2$) which transforms E in a one-to-one way into a nondegenerate cubic curve and transforms p into an inflection point of that curve. To see this, we again follow Tate's notes and assume that our given point p is not an inflection point. Choose coordinates so that

$$p = (1, 0, 0),$$

the tangent line to E at p is given by

$$z = 0,$$

and the tangent line to E at the third point of intersection of the line $z = 0$ with E is the line

$$x = 0.$$

Then the equation for E can be written

$$x^2 z + x Q(y, z) + C(y, z) = 0, \tag{2.10}$$

where

$$Q(y, z) = ay^2 + 2byz + cz^2$$

and

$$C(y, z) = syz^2 + tz^3.$$

Now we transform $\mathbb{QP}_2$ (and simultaneously $\mathbb{RP}_2$ and $\mathbb{CP}_2$) into itself by the rule

$$(x, y, z) \mapsto (xz, xy, z^2). \tag{2.11}$$

Let's stop a moment and visualize this transformation of $\mathbb{QP}_2$. First, it has an "inverse," namely

$$(x, y, z) \to (xy, y^2 z/x, yz) \tag{2.12}$$
$$\|$$
$$(x^2, yz, xz).$$

That is, the transformation (2.11) restricts to an automorphism of

$$(\mathbb{QP}_2 - \{xyz = 0\}).$$

The behavior of the map (2.11) on

$$xyz = 0$$

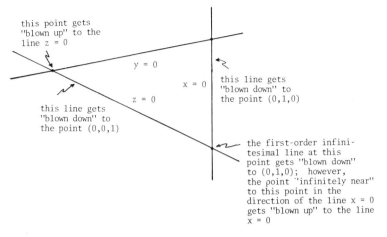

this point gets
"blown up" to the
line z = 0

y = 0

x = 0

z = 0

this line gets
"blown down" to
the point (0,0,1)

this line gets
"blown down" to
the point (0,1,0)

the first-order infini-
tesimal line at this
point gets "blown down"
to (0,1,0); however,
the point "infinitely near"
to this point in the
direction of the line x = 0
gets "blown up" to the line
x = 0

Figure 2.7. Analysis of the birational coordinate change (2.11).

can be pictured as shown in Figure 2.7. To describe what is happening more exactly, we consider the operation called "blowing up a point":

Suppose that we want to blow up the point

$$x = y = 0$$

in the Cartesian (x, y) plane. We simply replace the (x, y) plane with the graph of the map

$$\mathbb{Q}^2 \longrightarrow \mathbb{Q}\mathbb{P}_1,$$
$$(x, y) \longrightarrow (x, y)$$

in $\mathbb{Q}^2 \times \mathbb{Q}\mathbb{P}_1$.

With this in mind, it can be shown that the map (2.11) can be described in the following steps:

STEP 1: Blow up the points $(1, 0, 0)$ and $(0, 1, 0)$ in $\mathbb{Q}\mathbb{P}_2$ to obtain a new "manifold" X_1 (Figure 2.8).
STEP 2: Blow up the point p in X_1 to obtain a new manifold X_2 (Figure 2.9).

It can then be shown that there is an everywhere-defined "algebraic" mapping

$$f: X_2 \to \mathbb{Q}\mathbb{P}_2$$

which coincides with (2.11) on the part of X_2 and which was untouched

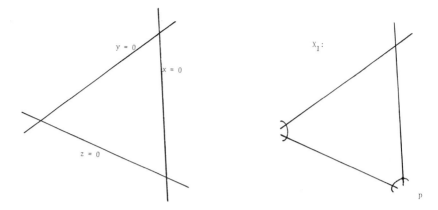

Figure 2.8. Pass from $\mathbb{QP}_2$ to X_1 by blowing up two points.

under the modifications of Steps 1 and 2. In fact, the net effect of f is to contract to points the sets indicated by the cross marks in Figure 2.10, in the order indicated by the number of cross marks.

The maps (2.11) and (2.12) are called *Cremona transformations*. The image of E under the transformation (2.11) is computed by replacing

$$x \text{ by } x^2,$$
$$y \text{ by } yz,$$
$$z \text{ by } xz$$

in the equation (2.10). We obtain

$$x^5z + x^2Q(yz, xz) + sx^2yz^3 + tx^3z^3 = 0,$$

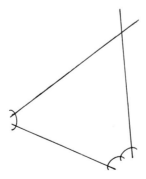

Figure 2.9. Pass from X_1 to X_2 by blowing up one more point.

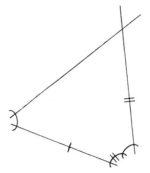

Figure 2.10. Pass from X_2 to $\mathbb{QP}_2$ by contracting three curves in the order shown.

that is,

$$x^2z[x^3 + ay^2z + 2bxyz + cx^2z + syz^2 + txz^2] = 0.$$

One sees easily that the part of E which is not being blown up or down is transformed nicely into the curve

$$-ay^2z - (2bx + sz)yz = x^3 + cx^2z + txz^2.$$

All this is more transparent if we write (2.10) in affine (x, y) coordinates:

$$x^2 + x(ay^2 + 2by + c) + sy + t = 0.$$

If we throw in an extra component (which will be collapsed to a point), the equation becomes

$$x^3 + ax^2y^2 + 2bx^2y + cx^2 + sxy + tx = 0.$$

The effect of the transformation (2.11) is to replace the quantity xy in this last equation by the quantity y, so we get

$$x^3 + ay^2 + 2bxy + cx^2 + sy + tx = 0.$$

Now replacing y by $[y - (bx/a + s/2a)]$, we eliminate the linear term in y in this last equation; and so by replacing (x, y) by $(c_1 x, c_2 y)$, we can reduce our equation to the form

$$y^2 = x^3 + Ax^2 + Bx + C. \tag{2.13}$$

If we follow the original point

$$p = (1, 0, 0)$$

through all these changes, we see that it gets blown up under (2.11) so that to calculate which point of the image cubic it corresponds to, we must look at the equation of E to second order at p. That is, what happens to the tangent line $z = 0$ to E at p under the transformation (2.11)? The fact that this whole line gets collapsed to the point $(0, 1, 0)$ allows us to conclude that the image of p is $(0, 1, 0)$. Since the remaining changes of coordinates leave the point at infinity $(0, 1, 0)$ fixed, our point p corresponds to the unique point at infinity of (2.13). Thus p does in fact go to an inflection point. If p is an inflection point to begin with, the transformation (2.11) is unnecessary in the normalization process.

In any case, if our original equation (2.9) has coefficients in $\mathbb{Q}$ (or $\mathbb{R}$), so does the final equation (2.13) and there is a nice bijection between (2.9) and (2.13) given in each direction by rational functions with rational coefficients. Thus we can find $\mathbb{Q}$ (or $\mathbb{R}$) points of (2.9) by finding those of (2.13). Also remember that for a second-degree equation with rational (real) coefficients, if one solution of the restriction of the equation to a rational (real) line is rational (real), so is the other. In the same way, if two solutions of the restriction of a cubic equation to a rational (real) line are rational (real), so is the third. Thus our geometric group law is well defined for cubics in $\mathbb{Q}\mathbb{P}_2$ and $\mathbb{R}\mathbb{P}_2$. A high point in the number theory of cubics is the theorem of Mordell, which says that the set of rational points of a rational cubic curve form a *finitely generated* abelian group.

2.5 A Thought about Complex Conjugation

The solution set of a nondegenerate real cubic curve is topologically either one or two circles, depending on whether the cubic polynomial in x in (2.13) has one or three real roots. To see how this solution set lies with respect to the set of complex solutions, we regard the latter as a two-sheeted covering of the x line (a *complex* line) ramified above the three points

$$p_1, p_2, p_3$$

and at ∞, where at least p_3 is real (Figure 2.11). Suppose now that p_1 and p_2 are not real. Then the real solution set to (2.13) projects onto a half-line connecting p_3 to infinity so that removing the real solution set from E simply makes it into a tube. The complex conjugation map

$$(x, y) \mapsto (\bar{x}, \bar{y})$$

leaves the real solution set pointwise fixed. If p_1 and p_2 are real, the removal of the two components of the real solution set from E breaks it up into two components which are interchanged under complex conjugation.

Figure 2.11. Branch points of the projection of a real cubic.

This last phenomenon carries over to nondegenerate curves of higher degree n whenever the solution set has the maximum number $[(n-1) \times (n-2)/2 + 1]$ of components.

2.6 Some Meromorphic Functions on Cubics

Let's look a bit more at stereographic projection from a point $p \in E$. This projection gives a map

$$E \longrightarrow \mathbb{CP}_1$$

or, what is the same thing, a meromorphic function on E. Given $p_1, p_2 \in E$, let p_3 be the third point of intersection of the line $\overline{p_1 p_2}$ with E. Then a sterographic projection with center p_3 can be used to construct a meromorphic function f on E whose only zeros are p_1 and p_2 and which has a pole at a pregiven point p_0 (see Figure 2.12). Suppose g is any meromorphic

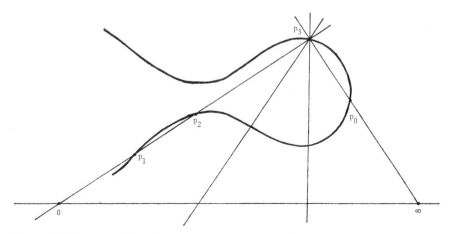

Figure 2.12. Stereographic projection from p_3 gives a meromorphic function with zero set $\{p_1, p_2\}$.

function whose zeros are p_1 and p_2 and which has a pole at p_0. Then the quotient

$$f/g$$

is zero at only one point (the other pole of g) and has only one simple pole (at the other pole of f). In other words, under the map

$$E \longrightarrow \mathbb{CP}_1,$$
$$p \longmapsto (f(p), g(p)),$$

0 and ∞ have unique preimages, so this map is either an isomorphism (which is a topological impossibility) or a constant. Thus all meromorphic functions of degree 2 on E arise via stereographic projection.

2.7 Cross Ratio Revisited, A Moduli Space for Cubics

We have seen that for any stereographic projection, the branch points, that is, the points on the image $\mathbb{CP}_1$ over which the map is ramified, are distinct. Let us assign to $p \in E$ the complex number

$$(p_1, p_2, p_3, p_4),$$

that is, the cross ratio (see Section 1.6) of the four branch points of the stereographic projection with center p. Since the cross ratio depends on the *order* in which the four points are taken, we do not obtain in this way a well-defined function on E. To fix this up, let $E^{(r)}$ denote the *rth symmetric product* of E, that is, the quotient of the Cartesian product E^r under the equivalence relation induced by permuting the order of the entries in the r-tuples. Then we do have a well-defined map from E into the fourfold symmetric product $E^{(4)}$ of E with itself which does not hit the diagonal locus of $E^{(4)}$. Now we take the *fibered product*, that is, we take the set X of all pairs $(p, (p_1, p_2, p_3, p_4))$ such that (p_1, p_2, p_3, p_4) is one of the cross ratios which occurs when projecting from p:

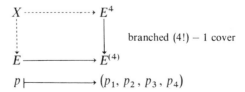

Then the cross-ratio map is well defined from X to $\mathbb{C}$. This function is holomorphic, and since the four points are always distinct, it never takes the values 0, 1, or ∞. So by the maximum principle it is constant. We have just seen that these sets of branch points are just the branch-point sets of degree-2 meromorphic functions on E, therefore we conclude:

> *If E and E' are cubic curves which are isomorphic as complex analytic manifolds, then the set of branch-point cross ratios for E must equal the set for E'.*

In fact, if any one cross ratio for E is equal to any one of those for E', then by elementary algebra the others must coincide as well, for if λ is one of the cross-ratios, there are only five other distinct ones, namely,

$$\frac{1}{\lambda}, \quad 1 - \lambda, \quad \frac{1}{1 - \lambda}, \quad \frac{\lambda - 1}{\lambda}, \quad \frac{\lambda}{\lambda - 1}.$$

Conversely, if E and E' have a common cross ratio λ, then an easy analytic continuation argument shows that each is isomorphic to the curve

$$y^2 = x(x - 1)(x - \lambda).$$

Thus we have constructed a moduli space, that is, a topological space whose points are in one-to-one correspondence with the set of isomorphism classes of cubic curves. This space is

$$\frac{\mathbb{C} - \{0, 1, \infty\}}{\text{equivalence relation}}$$

where $\lambda \sim \lambda'$ if λ' is any of the numbers $1/\lambda$, $1 - \lambda$, $1/(1 - \lambda)$, $(\lambda - 1)/\lambda$, $\lambda/(\lambda - 1)$.

2.8 The Abelian Differential on a Cubic

There is another rather deep connection between the analysis, geometry, and number theory of cubic curves. Suppose that we take our curve as given in the normal form

$$F(x, y) = y^2 - x(x - 1)(x - \lambda) = 0, \qquad \lambda \in \mathbb{Q}, \mathbb{R}, \text{ or } \mathbb{C}. \qquad (2.14)$$

Differentiating implicitly, we get

$$\left. \left(\frac{\partial F}{\partial x} \, dx + \frac{\partial F}{\partial y} \, dy \right) \right|_E = 0.$$

Now at the points $(0, 0)$, $(1, 0)$, and $(\lambda, 0)$ we have

$$\frac{\partial F}{\partial x} \neq 0,$$

which means (by the implicit function theorem) that y can be used as a local coordinate for E near those points. That is, if f is a holomorphic function on a neighborhood of one of these points in $\mathbb{CP}_2$, then $f|_E$ can be written locally as a power series in y. At other (finite) points of E,

$$x - x(\text{point in question})$$

can be used as a local coordinate. So

$$dy = \frac{-1}{2} \frac{\partial F/\partial x}{y} dx$$

is nonzero at the three points $(0, 0)$, $(1, 0)$, and $(\lambda, 0)$ and therefore so is

$$\frac{dx}{y}. \tag{2.15}$$

We also wish to check the behavior of the differential (2.15) at infinity, that is, at the point $(0, 1, 0)$ of the $\mathbb{CP}_2$ in which the (x, y) plane sits as an open dense subset. First we "homogenize" (2.14) to get

$$y^2 z - x(x - z)(x - \lambda z) = 0.$$

We then get the affine equation for E in the (x, z) plane [a neighborhood of $(0, 1, 0)$ in $\mathbb{CP}_2$] by setting $y = 1$. We obtain

$$G(x, z) = z - x(x - z)(x - \lambda z) = 0. \tag{2.16}$$

Since $\partial G/\partial z \neq 0$ at $(0, 0)$, the implicit function theorem again allows the use of x as a local coordinate for E there. So on E

$$z = \sum a_n x^n,$$

and substitution in (2.16) yields

$$a_0 = a_1 = a_2 = 0, \qquad a_3 = 1.$$

Now let's examine the differential (2.15). We homogenize it by writing it in the form

$$\frac{d(x/z)}{y/z} = \frac{x\,dz - z\,dx}{yz},$$

so on the (x, z) plane given by setting $y = 1$ the differential (2.15) becomes

$$(x/z)\,dz - dx = (2 + \text{positive powers of } x)\,dx. \tag{2.17}$$

Thus this differential is finite and nonzero at the infinite point of E. In the language of modern differential geometry, the section (2.15) of the holomorphic cotangent bundle of E has no zeros and no poles. Thus the cotangent bundle of E is trivial (and so the topological Euler characteristic is zero). For more information about this see, for example, [10, Section 41].

2.9 The Elliptic Integral

We pick $(0, 1, 0) \in E$ as a basepoint, which we call p_0, and we denote the differential (2.15) by ω. Consider the map

$$E \longrightarrow \mathbb{C}, \qquad (2.18)$$

$$p \longmapsto \int_{p_0}^{p} \omega = \int_{\infty}^{x(p)} \frac{dx}{[x(x-1)(x-\lambda)]^{1/2}}$$

This map is not well defined, of course, since we have not made precise our path of integration. Remember that we have realized E as a two-sheeted cover of the x axis cross-pasted along slits (see Figure 2.13). Also

$$\omega = [x(x-1)(x-\lambda)]^{-1/2} \, dx$$

has two determinations, depending on which value of the square root is picked. The differential ω on E is obtained by choosing one of these determinations, call it ω_1, for the top sheet, and the other, call it ω_2, for the bottom sheet. Then by the Cauchy integral formula it is clear that the mapping (2.18) is well defined modulo numbers of the form

$$m\pi_1 + n\pi_2, \qquad m, n \in \mathbb{Z},$$

where

$$\pi_1 = 2 \int_0^1 \omega_1, \qquad \pi_2 = 2 \int_1^\lambda \omega_1.$$

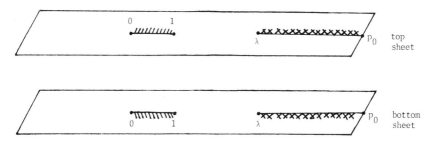

Figure 2.13. Floor plan of a cubic.

Also if we write $x = u + iv$ where u and v are real variables, then

$$\frac{i}{2}\int_{\mathbb{C}} \omega_1 \wedge \bar{\omega}_1 = \frac{i}{2}\int_{\mathbb{C}} \frac{(du + i\,dv) \wedge (du - i\,dv)}{|x(x-1)(x-\lambda)|}$$

$$= \int_{\mathbb{C}} \frac{1}{|x(x-1)(x-\lambda)|}\, du\, dv \qquad (2.19)$$

which, in polar coordinates near 0, is

$$\int \frac{1}{|x-1|\,|x-\lambda|}\, dr\, d\theta.$$

Making a similar development near 1, λ, and p_0, we see that this double integral over all of $\mathbb{C}$ is a finite positive number. On the other hand, if we write

$$\omega_i = \xi_i + i\eta_i, \qquad i = 1, 2,$$

where ξ_i and η_i are of the form

[real-valued function of (u, v)] du + [real-valued function of (u, v)] dv,

then

$$\int \frac{i}{2}\omega_1 \wedge \bar{\omega}_1 = \int \xi_1 \wedge \eta_1 > 0. \qquad (2.20)$$

We next wish to reduce this surface integral to a line integral. If we make a cut in the x axis as shown in Figure 2.14, then on the remainder of that complex line we can, by Green's theorem, find a function f such that

$$df = \xi_1 \quad \text{and} \quad \lim_{p \to \infty} f(p) = 0.$$

Again by Green's theorem

$$\int \xi_1 \wedge \eta_1 = \int_0^1 f^+ \eta_1^+ + \int_1^\lambda f^+ \eta_1^+ + \int_\lambda^\infty f^+ \eta_1^+$$

$$+ \int_1^0 f^- \eta_1^- + \int_\lambda^1 f^- \eta_1^- + \int_\infty^\lambda f^- \eta_1^-,$$

where $f1^+$ is the determination of f above the slit and f_1^- is its determination below the slit, etc.

$$0 \qquad 1 \qquad\qquad \lambda \qquad\qquad\qquad p_0 = \infty$$

Figure 2.14. Cutting the cubic into two slit complex number lines.

Now along the slit $\overline{\lambda\infty}$,

$$f^+ = -f^-;$$

along $\overline{1\lambda}$,

$$f^+ = f^- - 2\int_\lambda^\infty \xi_1^+ = f^- + 2\int_0^1 \xi_1^+ = f^- + \operatorname{Re}\pi_1;$$

along $\overline{01}$,

$$f^+ = -f^- + \operatorname{Re}\pi_1.$$

Also along $\overline{\lambda\infty}$,

$$\eta_1^+ = -\eta_1^-;$$

along $\overline{1\lambda}$,

$$\eta_1^+ = \eta_1^-;$$

and along $\overline{01}$,

$$\eta_1^+ = -\eta_1^-.$$

This all means that the integral (2.20) can be computed via Green's theorem:

$$
\int \xi_1 \wedge \eta_1 = \int_0^\infty f^+\eta_1^+ - \int_0^\infty f^-\eta_1^-
$$
$$
= \int_0^1 (\operatorname{Re}\pi_1)\eta_1^+ + \int_1^\lambda (\operatorname{Re}\pi_1)\eta_1^+
$$
$$
= \tfrac{1}{2}(\operatorname{Re}\pi_1)[\operatorname{Im}\pi_1 + \operatorname{Im}\pi_2].
$$

Repeating the same argument with

$$dg = \eta_1 \qquad \text{and} \qquad \lim_{p\to\infty} g(p) = 0,$$

we obtain

$$
\int \xi_1 \wedge \eta_1 = -(\operatorname{Im}\pi_1)\left[\int_0^1 \xi_1^+ + \int_1^\lambda \xi_1^+\right]
$$
$$
= \frac{-1}{2}(\operatorname{Im}\pi_1)[(\operatorname{Re}\pi_1) + (\operatorname{Re}\pi_2)].
$$

Adding gives

$$(\operatorname{Re}\pi_1)(\operatorname{Im}\pi_2) - (\operatorname{Im}\pi_1)(\operatorname{Re}\pi_2) > 0. \tag{2.21}$$

This means that the two complex numbers π_1 and π_2 are linearly independent over $\mathbb{R}$ so that

$$\tilde{E} = \frac{\mathbb{C}}{Z\pi_1 + Z\pi_2}$$

is a compact complex torus. The relation (2.21) is called the *second Riemann relation*.

Thus (2.18) gives a well-defined holomorphic immersion

$$E \to \tilde{E}. \tag{2.22}$$

Since E is compact, this mapping must be a finite covering space mapping. In fact, it is immediate to see that the mapping (2.22) induces an isomorphism

$$H_1(E; \mathbb{Z}) \cong H_1(\tilde{E}; \mathbb{Z})$$

or, what is the same, an isomorphism of fundamental groups. So by covering-space theory, the map (2.22) is itself an isomorphism

$$E \xrightarrow{\;\cong\;} \frac{\mathbb{C}}{\mathbb{Z}\pi_1 + \mathbb{Z}\pi_2}. \tag{2.23}$$

We shall later encounter the inverse isomorphism to (2.23). In particular, the composition of this inverse with the coordinate function x on E [see (2.18)] has special significance in the development of the subject, since it has a particularly nice explicit expression. It is called the *Weierstrass p-function*.

2.10 The Picard–Fuchs Equation

The same differential

$$\omega = \frac{dx}{y} = \frac{dx}{[x(x-1)(x-\lambda)]^{1/2}}$$

which gave the mapping (2.18) has another very surprising use. Namely, the numbers

$$\pi_1(\lambda) = 2 \int_0^1 \omega,$$

$$\pi_2(\lambda) = 2 \int_1^\lambda \omega, \tag{2.24}$$

called the *periods* of E, depend on λ. It is not hard to see that they are, in fact, holomorphic functions of λ. For instance $\pi_2(\lambda)$ can also be obtained by integrating ω around the path γ, as shown in Figure 2.15, so that as λ moves slightly, the path of integration does not change and we can differentiate $\pi_2(\lambda)$ by differentiation under the integral sign. Let

$$\pi(\lambda) = \pi_1(\lambda) + \pi_2(\lambda) = 2 \int_0^\lambda \omega.$$

$\pi(\lambda)$ can be obtained by integrating ω over the path (see Figure 2.16). As λ approaches 0, this integral becomes

$$\int_\gamma \frac{dx}{x(x-1)^{1/2}} = \text{residue}_0 \frac{1}{x(x-1)^{1/2}}$$

$$= 2\pi i(-i) = 2\pi.$$

Thus $\pi(\lambda)/2\pi$ has a power-series expansion around $\lambda = 0$. We wish to compute this expansion explicitly. To do this, we need to study the differential equation satisfied by the $\pi_i(\lambda)$, the so-called *Picard–Fuchs equation*.

And so we must compute the derivatives of $\pi_i(\lambda)$ with respect to λ; that is, we must differentiate under the integral sign; we must compute

$$\frac{\partial}{\partial\lambda}\left[x^{-1/2}(x-1)^{-1/2}(x-\lambda)^{-1/2}\right] = \tfrac{1}{2}x^{-1/2}(x-1)^{-1/2}(x-\lambda)^{-3/2},$$

$$\frac{\partial^2}{\partial\lambda^2}\left[x^{-1/2}(x-1)^{-1/2}(x-\lambda)^{-1/2}\right] = \tfrac{3}{4}x^{-1/2}(x-1)^{-1/2}(x-\lambda)^{-5/2}.$$

Now those readers familiar with the theory of deRham cohomology will recognize that there must be a relation between the three differentials

$$\omega, \ \frac{\partial\omega}{\partial\lambda}, \ \frac{\partial^2\omega}{\partial\lambda^2},$$

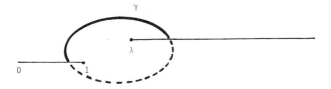

Figure 2.15. One of many equivalent paths of integration for $\pi_2(\lambda)$.

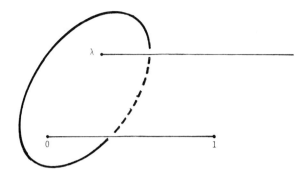

Figure 2.16. One of many equivalent paths of integration for $\pi(\lambda)$.

that is, some linear combination of these (with coefficients functions of λ) must be an exact differential. So the corresponding linear combination of

$$\pi_i, \ \frac{d\pi_i}{d\lambda}, \ \frac{d^2\pi_i}{d\lambda^2}$$

must be zero.

But let's just compute. If we fix λ and differentiate with respect to the variable x; we get

$$d\frac{x^{1/2}(x-1)^{1/2}(x-\lambda)^{1/2}}{(x-\lambda)^2} = \left[\frac{1}{2}x^{-1/2}(x-1)^{1/2}(x-\lambda)^{-3/2}\right.$$

$$+\frac{1}{2}x^{1/2}(x-1)^{-1/2}(x-\lambda)^{-3/2}$$

$$\left. -\frac{3}{2}x^{1/2}(x-1)^{1/2}(x-\lambda)^{-5/2}\right]dx$$

$$= (x-1)\frac{\partial\omega}{\partial\lambda} + x\frac{\partial\omega}{\partial\lambda} - 2x(x-1)\frac{\partial^2\omega}{\partial\lambda^2}$$

$$= [(x-\lambda)+(\lambda-1)]\frac{\partial\omega}{\partial\lambda} + [(x-\lambda)+\lambda]\frac{\partial\omega}{\partial\lambda}$$

$$-2[(x-\lambda)+\lambda][(x-\lambda)+(\lambda-1)]\frac{\partial^2\omega}{\partial\lambda^2}$$

$$= \frac{1}{2}\omega + (\lambda - 1)\ \frac{\partial\omega}{\partial\lambda} + \frac{1}{2}\omega + \lambda\ \frac{\partial\omega}{\partial\lambda} - 2[(x - \lambda) + \lambda]$$

$$\times \left(\frac{3}{2}\frac{\partial\omega}{\partial\lambda} + (\lambda - 1)\ \frac{\partial^2\omega}{\partial\lambda^2} \right) = \omega + (2\lambda - 1)\ \frac{\partial\omega}{\partial\lambda} - \frac{3}{2}\omega$$

$$- 3(\lambda - 1)\ \frac{\partial\omega}{\partial\lambda} - 3\lambda\ \frac{\partial\omega}{\partial\lambda} - 2\lambda(\lambda - 1)\ \frac{\partial^2\omega}{\partial\lambda^2}$$

$$= -\frac{1}{2}\omega - (4\lambda - 2)\ \frac{\partial\omega}{\partial\lambda} - 2\lambda(\lambda - 1)\ \frac{\partial^2\omega}{\partial\lambda^2}.$$

Integrating both sides around our cycles, we get the Picard–Fuchs equation:

$$0 = \frac{1}{4}\pi_i + (2\lambda - 1)\ \frac{d\pi_i}{d\lambda} + \lambda(\lambda - 1)\ \frac{d^2\pi_i}{d\lambda^2}. \tag{2.25}$$

Now we will appeal to the elementary theory of ordinary differential equations with regular singular points.[†] The indicial polynomial of (2.25) is

$$q(r) = r(r - 1) + r = r^2.$$

This means that the vector space of solutions of (2.25) near $\lambda = 0$ is generated by

$\sigma_1(\lambda)$ holomorphic and nonvanishing at 0,

$(\lambda\sigma_2(\lambda) + (\log \lambda)\sigma_1(\lambda))$ where σ_2 is holomorphic and nonvanishing at 0.

Thus it must be that if we normalize so that

$$\sigma_1(0) = 1,$$

then

$$\sigma_1(\lambda) = \frac{\pi(\lambda)}{2\pi}.$$

[†] See E. Coddington, *An Introduction to Ordinary Differential Equations* (Englewood Cliffs, N.J.: Prentice-Hall, 1961), Chapter Four, for more details.

Let's solve explicitly for the power-series expansion

$$\sum_{n=0}^{\infty} a_n \lambda^n$$

of σ_1. First,

$$\sigma_1'(\lambda) = \sum_{n \geq 0} (n + 1)a_{n+1} \lambda^n,$$

$$\sigma_1''(\lambda) = \sum_{n \geq 0} (n + 2)(n + 1)a_{n+2} \lambda^n.$$

So from (2.25) we obtain

$$\sum_{n \geq 0} [\lambda(\lambda - 1)(n + 2)(n + 1)a_{n+2} + (2\lambda - 1)(n + 1)a_{n+1} + \tfrac{1}{4}a_n]\lambda^n = 0.$$

Rewriting we get

$$\sum_{n \geq 0} [(n + \tfrac{1}{2})^2 a_n - (n + 1)^2 a_{n+1}]\lambda^n = 0.$$

Since $a_0 = 1$ and we now know that

$$a_{n+1} = \left[\frac{n + (1/2)}{n + 1} \right]^2 a_n,$$

we obtain

$$a_n = \binom{-1/2}{n}^2. \tag{2.26}$$

Notice that we have obtained along the way that near $\lambda = 0$

$$\pi_1(\lambda) \sim \log \lambda.$$

To sum up, we have a very specific power series

$$\sum_{n=0}^{\infty} \binom{-1/2}{n}^2 \lambda^n,$$

which gives us "the" solution to the Picard–Fuchs equation that stays bounded near the singular point $\lambda = 0$. It is no accident that the coefficients in this power series are rational, as we shall soon see.

2.11 Rational Points on Cubics Over $\mathbb{F}_p$

So far everything we have done with the differential ω and the function $\pi(\lambda)$ lies in the realm of analysis or geometry. But surprisingly enough these computations have number-theoretic applications. For this we must consider cubic curves over the finite fields

$$\mathbb{F}_p = \mathbb{Z}/p\mathbb{Z},$$

where p is an odd prime number. Given $\lambda \in \mathbb{Z}$, we reduce λ mod p and let C_λ denote the solution set of

$$y^2 = x(x - 1)(x - \lambda)$$

in $\mathbb{F}_p \times \mathbb{F}_p$. The question we will ask is,

What is the cardinality of C_λ?

Or, equivalently,

For how many $x \in \mathbb{F}_p$ is it true that $x(x - 1)(x - \lambda) \in \mathbb{F}_p^2$?

Now since the multiplicative group $(\mathbb{F}_p - \{0\})$ is cyclic, $x(x - 1) \times (x - \lambda) \in \mathbb{F}_p^2$ if and only if

$$[x(x - 1)(x - \lambda)]^{(p - 1)/2} \equiv 1 \quad \text{or} \quad x = 0, 1, \lambda.$$

In all other cases $[x(x - 1)(x - \lambda)]^{(p - 1)/2} \equiv -1$. We can summarize all this in the very neat formula

$$(\text{number of points in } C_\lambda) \equiv \sum_{x \in \mathbb{F}_p} \{1 + [x(x - 1)(x - \lambda)]^{(p - 1)/2}\}$$

(2.27)

modulo p. We next wish to simplify the right-hand side of (2.27). For this we use the (character) formulas

$$\sum_{x \in \mathbb{F}_p} x^k \equiv 0 \qquad \text{if } (p - 1) \nmid k,$$

$$\sum_{x \in \mathbb{F}_p} x^k \equiv -1 \qquad \text{if } (p - 1) \mid k.$$

(2.28)

These are easily proved by noting that

$$\sum x^k = y^k \sum x^k$$

for any $y \neq 0$. Now suppose we write

$$[x(x - 1)(x - \lambda)]^{(p - 1)/2}$$

(2.29)

as a polynomial in x. Then by (2.28) the only term in (2.29) which will contribute to

$$\sum_{x \in \mathbb{F}_p} [x(x - 1)(x - \lambda)]^{(p - 1)/2}$$

is the term involving x^{p-1}. The coefficient of this term is the coefficient of $x^{(p-1)/2}$ in the polynomial expansion of

$$[(x-1)(x-\lambda)]^{(p-1)/2}.$$

But we can multiply this last expression out explicitly:

$$\left[\sum_{k=0}^{(p-1)/2}\binom{(p-1)/2}{k}(-1)^k x^{(p-1)/2-k}\right]\left[\sum_{l=0}^{(p-1)/2}\binom{(p-1)/2}{l}(-1)^l \lambda^l x^{(p-1)/2-l}\right]$$

has as coefficient of $x^{(p-1)/2}$ the sum

$$(-1)^{(p-1)/2}\sum_{k+l=(p-1)/2}\binom{(p-1)/2}{k}\binom{(p-1)/2}{l}\lambda^l$$

$$= (-1)^{(p-1)/2}\sum_{r=0}^{(p-1)/2}\binom{(p-1)/2}{r}^2 \lambda^r$$

$$\equiv (-1)^{(p-1)/2}\sum_{r=0}^{(p-1)/2}\binom{-1/2}{r}^2 \lambda^r$$

modulo p. This last congruence needs a little explanation. The integer

$$\frac{1}{r!}\cdot\frac{p-1}{2}\cdot\frac{p-3}{2}\cdot\ \cdots\ \cdot\frac{p-(2r-1)}{2}$$

represents the same element of $\mathbb{F}_p$ as does

$$a(p-1)(p-3)\cdot\ \cdots\ \cdot[p-(2r-1)], \tag{2.30}$$

where a is any integer such that

$$ar!2^r \equiv 1 \mod p.$$

But then

$$(2.30) \equiv a\cdot(-1)(-3)\cdot\ \cdots\ \cdot[-(2r-1)] \mod p,$$

which gives the desired congruence.

Now if $r \geq (p+1)/2$, then

$$\binom{-1/2}{r} = \frac{(-1)(-3)\cdots(-p)\cdots}{r!2^r} \equiv 0,$$

so the formula for the cardinality modulo p of C_λ is

$$(-1)(-1)^{(p-1)/2}\sum_{r=0}^{\infty}\binom{-1/2}{r}^2 \lambda^r.$$

This is the *same* formula as the one for the period function $\pi(\lambda)$, the solution to the Picard–Fuchs equation which is holomorphic at 0! Of course, this fact is not accidental—the reason for it is a rather deep one, discovered only in recent years by Y. Manin. To close this chapter, we will try to give some idea of Manin's result.

2.12 Manin's Result: The Unity of Mathematics

The fundamental ingredient in this discussion will be the algebra-geometric version of the Lefschetz fixed-point theorem from topology. The topological version says that if f is a differentiable mapping

$$f : M \rightarrow M,$$

where M is a compact differentiable manifold such that the graph of f meets the diagonal transversely in $M \times M$, then the Lefschetz number

$$L(f) = \sum_{p \in M} \sigma_p(f) = \sum_{n=0}^{\infty} (-1)^n \, \mathrm{trace}[f^* : H^n(M; \mathbb{C}) \rightarrow H^n(M; \mathbb{C})],$$

where

$$\sigma_p = \begin{cases} 0 & \text{if } f(p) \neq p, \\ \begin{matrix} +1 \\ -1 \end{matrix} & \text{if (graph } f) \text{ meets (diagonal) with} \begin{matrix} |\text{positive}| \\ |\text{negative}| \end{matrix} \text{orientation.}\dagger \end{cases}$$

Now transversality at p means, in local coordinates, that the mapping

$$(\text{identity} - f)$$

has maximal rank at p. So

$$\sigma(p) = \mathrm{sign} \det[I - J_p(f)],$$

where $J_p(f)$ is the Jacobian matrix of f at p. Now if A is a complex-valued matrix in triangular form

$$A = \begin{bmatrix} a_1 & & \\ & \ddots & * \\ 0 & & a_n \end{bmatrix},$$

then

$$\det[I - A] = \sum_{r=0}^{n} (-1)^r \sum_{j_1 < \cdots < j_r} a_{j_1} \cdot \cdots \cdot a_{j_r} = \sum_{r=0}^{n} (-1)^r \, \mathrm{trace}(\wedge^r A),$$

where $\wedge^r A$ denotes the linear endomorphism which A induces on the rth

---

† See, for example, Section 7, Chapter Four, of E. Spanier's *Algebraic Topology* (New York: McGraw-Hill, 1966).

exterior power of $\mathbb{C}^n$. Thus we can rewrite the Lefschetz fixed-point formula as

$$L(f) = \sum_{r,\,p} (-1)^r \frac{\text{trace } \Lambda^r J_p(f)}{|\det[1 - J_p(f)]|} = \sum_{n=0}^{\infty} (-1)^n \text{trace}(f^*|_{H^n(M)}). \quad (2.31)$$

The main point to notice is that we have two formulas for $L(f)$, one given by *local* invariants and the other by *global* ones.

Now if M is a compact complex manifold, let Γ be the global sections functor, and let

$$\Gamma \mathscr{A}^{0,0} \xrightarrow{\ \bar{\partial}\ } \Gamma \mathscr{A}^{0,1} \xrightarrow{\ \bar{\partial}\ } \Gamma \mathscr{A}^{0,2} \xrightarrow{\ \bar{\partial}\ } \cdots \quad (2.32)$$

be the Dolbeault complex on M, that is

$$\mathscr{A}^{0,q} = \text{sheaf of } C^\infty \ (0, q)\text{-forms on } M.$$

(For more details on this, see Gunning and Rossi [2].) The cohomology groups of this complex are denoted

$$H^q(M; \mathcal{O}) \quad \text{or} \quad H^{0,q}(M).$$

If M is a Kähler manifold, then $H^q(M; \mathcal{O})$ is a direct summand of the complex-valued deRham cohomology. Suppose now our mapping

$$F: M \to M$$

is holomorphic. Then f induces a morphism on the sequence (2.32), and the formula (2.31) continues to hold when the deRham complex is replaced by the complex (2.32). In that case formula (2.31) reads

$$\sum_{r,\,p} (-1)^r \frac{\text{trace } \Lambda^r J_p''(f)}{|\det[1 - J_p(f)]|} = \sum_{n=0}^{\infty} (-1)^n \text{trace}(f^*|_{H^n(M,\,\mathcal{O})}),$$

where $J_p''(f)$ is the restriction of the cotangent space mapping $J_p(f)$ to the subspace of type $(0, 1)$. But as before

$$\sum_r (-1)^r \text{trace } \Lambda^r J_p''(f) = \det[1 - J_p''(f)],$$

and if $J_p'(f)$ denotes the $(1, 0)$-part of $J_p(f)$, then

$$\det[1 - J_p(f)] = \det[1 - J_p'(f)] \cdot \det[1 - J_p''(f)].$$

Thus our formula becomes

$$\sum (-1)^n \text{trace}(f^*|_{H^n(M,\,\mathcal{O})}) = \sum_{p \text{ fixed}} \frac{1}{\det[1 - J_p'(f)]}. \quad (2.33)$$

Now the marvelous fact is that formula (2.33) continues to hold in a purely algebraic context. The reader who is unfamiliar with sheaf cohomology in algebraic geometry over an arbitrary algebraically closed field k is urged to read on in any case. The results used will be the formal analogues of the corresponding theorems over $\mathbb{C}$, and that formal analogy will probably carry the reader through. In any case, the beauty of the result should motivate most interested readers to further study of these topics.

Suppose, for instance, that k is the algebraic closure of the field $\mathbb{F}_p$ of p elements, $\lambda \in (\mathbb{F}_p - \{0, 1\})$, and M is the solution set to

$$y^2 z = x(x - z)(x - \lambda z)$$

in kP_2 (one-dimensional subspaces of k^3). In other words, M is the solution set to

$$y^2 = x(x - 1)(x - \lambda)$$

in k^2 with the point at infinity thrown in. Let f be the Frobenius mapping

$$f: \quad M \longrightarrow M$$
$$(x, y, z) \longrightarrow (x^p, y^p, z^p).$$

Now $d(x^p)/dx = px^{p-1} = 0$, and

$$H^n(M; \mathcal{O}) = 0, \qquad \text{if } n > 1 = \dim_k M$$

for the *sheaf* $\mathcal{O}$ of regular algebraic functions on M. Thus the formula (2.33) reads

$$1 - \text{trace}(f^*|_{H^1(M;\mathcal{O})}) = \text{number of fixed points of } f. \qquad (2.34)$$

But the fixed points of f are nothing more than the number of points of M which are represented by triples (x, y, z) of elements of $\mathbb{F}_p$, since

$$x = x^p \qquad \text{if and only if } x \in \mathbb{F}_p.$$

Since the point at infinity is one fixed point of f, the formula (2.34) becomes

$$-\text{trace } f^*|_{H^1(M;\mathcal{O})} = \text{number of points in } C_\lambda.$$

So we will explain the connection between the formula for the cardinality of C_λ and the period function $\pi(\lambda)$ by explicitly computing the trace of f^* on $H^1(M; \mathcal{O})$. We will have to use the fact that if $q, q' \in M$, then $H^1(M; \mathcal{O}) \cong$

$$\frac{\text{(algebraic functions with poles at } q \text{ and } q')}{\text{(functions with poles at } q) + \text{(functions with poles at } q')}. \qquad (2.35)$$

This comes out of the exact sheaf sequence

$$0 \to \mathcal{O} \to \mathcal{O}(\infty \cdot q) + \mathcal{O}(\infty \cdot q') \to \mathcal{O}(\infty \cdot q + \infty \cdot q') \to 0,$$

where

$\mathcal{O}(n \cdot q)$ = sheaf of algebraic functions on M with q as the only pole and no worse than order $(-n)$ at q,

$$\mathcal{O}(\infty \cdot q) = \lim_{n \to \infty} \mathcal{O}(n \cdot q).$$

Also, using formal power series, we can define differentials on M, and as in the complex case, the k-vector space of everywhere-regular differentials is one dimensional with generator written locally at q

$$\omega = dx + \sum_{r \geq 1} a_r(\lambda)[x - x(q)]^r \, dx. \tag{2.36}$$

Also just as before,

$$1 + \sum_{r \geq 1} a_r(\lambda)[x - x(q)]^r$$

satisfies the Picard–Fuchs equation, that is,

$$\left(\lambda(\lambda - 1)\frac{\partial^2}{\partial \lambda^2} + (2\lambda - 1)\frac{\partial}{\partial \lambda} + \frac{1}{4} \right)\left(1 + \sum_{r \geq 1} a_r(\lambda)[x - x(q)]^r \right)$$
$$= \frac{d}{dx}\left(\text{series expansion for } \frac{x^{1/2}(x - 1)^{1/2}(x - \lambda)^{1/2}}{(x - \lambda)^2} \right). \tag{2.37}$$

Furthermore, referring again to (2.35), we see that the pairing

$$H^1(M; \mathcal{O}) \quad \times \quad \left| \begin{matrix} \text{regular} \\ \text{differentials} \end{matrix} \right| \quad \longrightarrow \quad k \tag{2.38}$$
$$h \qquad\qquad \times \qquad\qquad \omega \qquad\qquad \longmapsto \quad \text{res}_q(h\omega)$$

is nondegenerate, so, in particular,

$$\dim_k H^1(M; \mathcal{O}) = 1.$$

This is the algebraic version of the *Serre duality theorem*. Using these facts we are in a position to compute

$$\text{trace}[f^* : H^1(M; \mathcal{O}) \to H^1(M; \mathcal{O})].$$

Namely, the algebraic *Riemann–Roch theorem* (see Chapter Three) will always assure us of the existence of an algebraic function h whose only poles are q and q' and which has a simple pole at q. We write

$$h = \frac{1}{x - x(q)} + \sum_{l \geq 0} b_l[x - x(q)]^l.$$

Then the map f sends $h(x)$ to

$$h(x^p) = \frac{1}{[x - x(q)]^p} + \sum_{l \geq 0} b_l[x - x(q)]^{pl}$$

as long as q is chosen to lie in C_λ. So the trace of f^* must be the coefficient of

$$\frac{1}{[x - x(q)]}$$

in the series expansion of $h(x^p)\omega$. That is, referring to (2.36), we have

$$(\text{trace } f^*) = a_{p-1}(\lambda). \tag{2.39}$$

But now by (2.37)

$$\left(\lambda(\lambda - 1)\frac{\partial^2}{\partial \lambda^2} + (2\lambda - 1)\frac{\partial}{\partial \lambda} + \frac{1}{4}\right)a_{p-1}(\lambda)[x - x(q)]^{p-1}$$

$$= \frac{d}{dx}\{c(\lambda)[x - x(q)]^p\} = 0,$$

so that $a_{p-1}(\lambda)$ *satisfies the Picard–Fuchs equation!* Also $a_{p-1}(\lambda)$ is univalued around $\lambda = 0$, so to compute its series expansion we make the formal computations leading to (2.26), and we obtain

$$a_{p-1}(\lambda) = c \sum_{r=0}^{\infty} \binom{-1/2}{r}^2 \lambda^r.$$

To evaluate c then we need only compute the number of points of C_λ for one value of λ (which we have already done). Thus

$$\text{trace } f^*|_{H^1(M;\, \mathcal{O})} = -\text{cardinality of } C_\lambda$$

$$= (-1)^{(p-1)/2} \sum_{r=0}^{\infty} \binom{-1/2}{r}^2 \lambda^r$$

$$= (-1)^{(p-1)/2} \sum_{r=0}^{(p-1)/2} \binom{-1/2}{r}^2 \lambda^r.$$

2.13 Some Remarks on Serre Duality

The usual form of Serre duality in the theory of complex manifolds is not the one we used to make this last computation. Instead we used the nondegeneracy of the pairing

$$H^i(M; \Omega^j) \otimes H^k(M; \Omega^l) \to H^m(M; \Omega^m), \tag{2.40}$$

where $(i + k) = (j + l) = m =$ complex dimension of the complex manifold M, and

$$\Omega^l = \text{sheaf of holomorphic } l \text{ forms on } M.$$

$H^*(M; \Omega^j)$ is computed from the exact sequence

$$\Gamma(\Omega^j \otimes \mathscr{A}^{0,\,0}) \xrightarrow{\ \bar{\partial}\ } \Gamma(\Omega^j \otimes \mathscr{A}^{0,\,1}) \xrightarrow{\ \bar{\partial}\ } \cdots,$$

which generalizes (2.32). The pairing (2.40) is given by

$$(\omega, \eta) \to \int_M \omega \wedge \eta,$$

where ω is a $\bar{\partial}$-closed (j, i) form and η is a $\bar{\partial}$-closed (l, k) form. Thus in the case which concerns us, the usual complex version reads

$$H^1(M; \mathcal{O}) \otimes H^0(M; \Omega^1) \longrightarrow \mathbb{C}, \qquad (2.41)$$

$$(\eta, \omega) \longmapsto \int_M \eta \wedge \omega.$$

To see that this is the same (in the complex case) as the pairing (2.38), consider the open covering

$$U_1 = (M - q), \qquad U_2 = (M - q')$$

of M. Then if f is a function with simple pole at q and only other pole q', the corresponding Čech one-cochain with coefficients in $\mathcal{O}$ is given by

$$f|_{U_1 \cap U_2} \in C^1(M; \mathcal{O}). \qquad (2.42)$$

Next construct the double complex of Čech groups

$$
\begin{array}{ccccccc}
{\scriptstyle f|_{U_1 \cap U_2}\uparrow\delta} & & {\scriptstyle f|_{U_1 \cap U_2}\uparrow\delta} & & & {\scriptstyle \uparrow\partial} & \\
0 \longrightarrow C^1(M; \mathcal{O}) \longrightarrow & C^1(M; \mathscr{A}^{0,\,0}) & \xrightarrow{\bar{\partial}} & C^1(M; \mathscr{A}^{0,\,1}) & \xrightarrow{\bar{\partial}} & & \\
{\scriptstyle \uparrow\delta} & {\scriptstyle (g,\,h)\uparrow\delta} & & {\scriptstyle (\bar{\partial}g,\,\bar{\partial}g)\uparrow\delta} & & & \\
0 \longrightarrow C^0(M; \mathcal{O}) \longrightarrow & C^0(M; \mathscr{A}^{0,\,0}) & \longrightarrow & C^0(M; \mathscr{A}^{0,\,1}) & \xrightarrow{\bar{\partial}} & & \\
{\scriptstyle \uparrow} & {\scriptstyle \uparrow} & & {\scriptstyle (\bar{\partial}g)\uparrow} & & & \\
0 \qquad \mathbb{C} \longrightarrow & \Gamma(\mathscr{A}^{0,\,0}) & \xrightarrow{\bar{\partial}} & \Gamma(\mathscr{A}^{0,\,1}) & \xrightarrow{\bar{\partial}} & & \\
{\scriptstyle \uparrow} & {\scriptstyle \uparrow} & & {\scriptstyle \uparrow} & & & \\
0 & 0 & & 0 & & &
\end{array}
$$

and follow the cochain (2.42) across to the corresponding representative $\bar{\partial}g$ in $\Gamma\mathscr{A}^{0,1}$. To do this, pick

$$g\begin{vmatrix} \text{smooth at } q', \\ = (1/2)f \text{ away from a neighborhood of } q', \end{vmatrix}$$

$$h\begin{vmatrix} \text{smooth at } q, \\ = (-1/2)f \text{ away from a neighborhood of } q, \end{vmatrix}$$

such that $g - h = f$ everywhere except at the points q and q'. Then

$$\bar{\partial}g = \bar{\partial}h,$$

and the equivalence of the pairings (2.38) and (2.41) follows from the equation

$$\int_q f\omega = \lim_{\text{radius}\to 0} \int_q g\omega = \int_M \omega \wedge dg = \int_M \omega \wedge \bar{\partial}g.$$

The point is that the (equivalent) form (2.38) of the pairing carries over to the general algebraic case whereas the form (2.41) of the pairing makes no sense except when we are working over the complex numbers.

CHAPTER THREE

Theta Functions

3.1 Back to the Group Law on Cubics

In this chapter we will attack the problem of giving a nice structure to the set of isomorphism classes of cubics and also the set of "framed" cubics. We rely heavily on the analytic tools of theta functions and modular forms. The introduction of these concepts will also prepare us for their use in a more general setting in later chapters.†

In Section 2.9 we saw that if E is a nonsingular cubic in $\mathbb{C}\mathbb{P}_2$, and if γ_1 and γ_2 form a basis for

$$H_1(E; \mathbb{Z})$$

and

$$\pi_j = \int_{\gamma_j} \omega,$$

then there is an isomorphism of analytic sets

$$f: \quad E \longrightarrow \frac{\mathbb{C}}{\mathbb{Z}\pi_1 + \mathbb{Z}\pi_2} = \tilde{E} \qquad (3.1)$$

$$p \longrightarrow \int_{p_0}^{p} \omega.$$

---

† An algebraic version of the material of this chapter, which extends many of the results to cubics in characteristic p, can be found in "Algebraic Formulas in Arbitrary Characteristic," by John Tate, which appears as Appendix 1 in S. Lang's book, *Elliptic Functions* (Reading, Mass: Addison-Wesley, 1973).

Since both E and $\tilde{E}$ are groups, we would like this map to be a group homomorphism. To see that this is indeed the case, let $\mathbb{CP}_2^*$ denote the set of lines in $\mathbb{CP}_2$ and define

$$
\begin{array}{ccc}
\mathbb{CP}_2^* & \longrightarrow & \tilde{E}, \\
L & \longmapsto & \displaystyle\sum_{p \in (L \cap E)} f(p).
\end{array}
\tag{3.2}
$$

This mapping is easily seen to be holomorphic and everywhere defined. Since $\mathbb{CP}_2^*$ is simply connected, the mapping (3.2) lifts to a mapping

$$
\mathbb{CP}_2^* \to \mathbb{C},
$$

since $\mathbb{C}$ is the universal covering space of $\tilde{E}$. This last map must be a constant by the maximum principle, and since the line

$$
z = 0
$$

must go to $0 \in \tilde{E}$ by the way we normalized E, we have that

$$
\begin{aligned}
& p_1 + p_2 + p_3 = 0 \quad \text{in } E \quad \text{if and only if} \\
& \{p_1, p_2, p_3\} = L \cap E \text{ for } L \in \mathbb{CP}_2^*,
\end{aligned}
$$

which implies

$$
f(p_1) + f(p_2) + f(p_3) = 0 \qquad \text{in } \tilde{E}.
$$

Thus (3.1) is a group isomorphism as well.

We should insert two remarks at this point. First, suppose we have a complex analytic mapping

$$
f: \frac{\mathbb{C}}{\pi_1 \mathbb{Z} + \pi_2 \mathbb{Z}} \to \frac{\mathbb{C}}{\pi_1' \mathbb{Z} + \pi_2' \mathbb{Z}}.
\tag{3.3}
$$

Then the function df/dz is doubly periodic and so is constant. This means that if we couple f with an appropriate translation and lift to a transformation

$$
\mathbb{C} \to \mathbb{C},
$$

the resulting map is simply

$$
z \mapsto cz
$$

for some *constant* c. Thus:

Up to translation, f is automatically a group homomorphism.

Second, it is natural to ask whether *all* complex manifolds of the form

$$\frac{\mathbb{C}}{(\pi_1\mathbb{Z} + \pi_2\mathbb{Z})},$$

the so-called *elliptic curves*, come from cubic curves by the process we have outlined. We shall soon see that this is indeed the case. However, before doing this, we wish to settle another question which we touched on in Chapters One and Two.

3.2 You Can't Parametrize a Smooth Cubic Algebraically

We were able to "parametrize" conics by using the inverse of stereographic projection to give a mapping

$$\mathbb{C} \to (\text{conic}), \tag{3.4}$$

and now we use the "inverse" of the multivalued map

$$E \xrightarrow{\hspace{2cm}} \mathbb{C}$$

$$p \xrightarrow{\hspace{1.5cm}} \int_0^p \omega$$

to give a parametrization

$$\mathbb{C} \to E. \tag{3.5}$$

Later we shall give the mapping (3.5) more explicitly, but for now let's just compare it with (3.4).

The mapping (3.4) is of course algebraic (given by rational functions) whereas the mapping (3.5) is not. In fact, there is no quotient of polynomials

$$f(z) = p_1(z)/p_2(z), \qquad g(z) = q_1(z)/q_2(z)$$

such that

$$g^2 = f(f-1)(f-\lambda) \tag{3.6}$$

except when f and g are constant functions. For if (3.6) were true, then

$$p_2^3 q_1^2 = q_2^2 p_1(p_1 - p_2)(p_1 - \lambda p_2).$$

Then since we can (and will) assume that p_1 and p_2 are relatively prime elements of the polynomial ring $\mathbb{C}[z]$ and that q_1 and q_2 are also relatively prime, we have that

$$p_2^3 \mid q_2^2 \quad \text{and} \quad q_2^2 \mid p_2^3.$$

Thus adjusting q_1 by an appropriate constant gives

$$q_1^2 = p_1(p_1 - p_2)(p_1 - \lambda p_2). \tag{3.7}$$

Also since $p_2^3 = (\text{const})q_2^2$, p_2 is a perfect square. So now consider the family of polynomials

$$t_1 p_1 + t_2 p_2, \qquad (t_1, t_2) \in \mathbb{CP}_1.$$

By (3.7) this family has four distinct entries given by

$$(t_1, t_2) = (1, 0), (1, -1), (1, -\lambda), (0, 1),$$

which are perfect squares. But now if

$$ar_1^2 - br_2^2 = \text{square},$$
$$cr_1^2 - dr_2^2 = \text{square}$$

for some relatively prime polynomials r_1 and r_2, we have

$$a^{1/2}r_1 - b^{1/2}r_2 = \text{square},$$
$$a^{1/2}r_1 + b^{1/2}r_2 = \text{square},$$
$$c^{1/2}r_1 - d^{1/2}r_2 = \text{square},$$
$$c^{1/2}r_1 + d^{1/2}r_2 = \text{square},$$

so that the family of polynomials

$$t_1 r_1 + t_2 r_2$$

becomes a square for four distinct values of (t_1, t_2) in $\mathbb{CP}_1$. Repeating the argument and noting that the degree of the polynomials in question is halved at each step, we arrive at a contradiction.

Notice that the preceding argument depended on the fact that

$$\lambda \neq 0, 1$$

If, for example, $\lambda = 0$, then we do have a rational parametrization of the *singular* curve E. This is achieved by means of stereographic projection with center the singular point $(0, 0)$, for every line through $(0, 0)$ meets E in only one more point, just as happened in the case of smooth conics (see Figure 3.1). If the line is given by

$$y = \alpha x,$$

and we parametrize the line by x, then the third point of intersection is given by

$$\alpha^2 x^2 = x^2(x - 1),$$

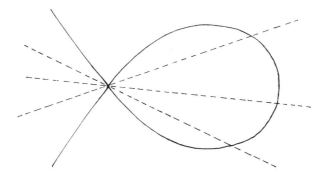

Figure 3.1. Stereographic projection of a singular cubic.

that is,

$$x = \alpha^2 + 1$$

So our third point of intersection is simply

$$(\alpha^2 + 1, \alpha^3 + \alpha).$$

Now if $\alpha = \pm i$, this point is $(0, 0)$. The corresponding lines are the two lines with contact of order 3 with E at $(0, 0)$. (See Figure 3.2.) Together they make up the *tangent cone* of E at $(0, 0)$, and their equation is obtained by simply eliminating from the equation

$$y^2 = x^2(x - 1)$$

all terms which are not of lowest degree, to get

$$y^2 = -x^2.$$

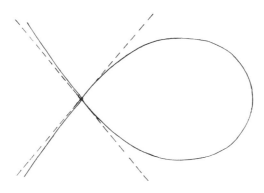

Figure 3.2. The singular point on the cubic has two tangent lines.

So stereographic projection from $(0, 0)$ is given by

$$
\begin{array}{rl}
E & \longrightarrow \mathbb{CP}_1, \\
(x, y) & \longmapsto (y, x),
\end{array}
\tag{3.8}
$$

which is a bijection except that it "splits" the point $(0, 0)$, that is:

> The point $(0, 0)$ with the "infinitely near point" (i.e., tangent direction) $\alpha = i$ goes to $(i, 1)$.

> The point $(0, 0)$ with the "infinitely near point" (i.e., tangent direction) $\alpha = -i$ goes to $(-i, 1)$.

We say that $\mathbb{CP}_1$ is the *desingularization* of E. The map built from the inverse of (3.8) becomes an everywhere-defined algebraic map which is onto E and one-to-one except that $(0, 0)$ has two preimages:

$$
\begin{array}{rl}
\mathbb{CP}_1 & \longrightarrow E, \\
(x, y) & \longmapsto \left((x/y)^2 + 1, \, (x/y)((x/y)^2 + 1) \right).
\end{array}
$$

3.3 Meromorphic Functions on Elliptic Curves

Next let us examine the field of meromorphic functions on a nonsingular cubic curve.

We want to build functions on E or, what is the same, doubly periodic functions on $\mathbb{C}$ with periods π_1 and π_2 (linearly independent over $\mathbb{R}$). Replacing $\omega = dx/y$ by $c\omega$ for appropriate nonzero constant c, we can assume that

$$
\pi_1 = 1.
\tag{3.9}
$$

Then the second Riemann relation, which we saw in Section 2.9, says that

$$
(\operatorname{Im} \pi_2) > 0.
\tag{3.10}
$$

From now on we will write τ instead of π_2. Then if f is periodic with period 1 and is holomorphic, it has a Fourier expansion

$$
f(u) = \sum_{n=-\infty}^{\infty} a_n e^{2\pi i n u}.
$$

If

$$
f(u + \tau) = f(u),
$$

we would have the contradiction that

$$
a_n e^{2\pi i n \tau} = a_n.
\tag{3.11}
$$

The closest we can come to this is to demand a second periodicity property, which results in a relation between a_n and a_{n+1} to replace the relation (3.11). The set of such functions will then form (at most) a one-dimensional vector space. We accomplish this by demanding that

$$f(u + \tau) = e^{-2\pi i(u + \alpha)} f(u).$$

This translates to the identity

$$\sum a_n e^{2\pi in\tau} e^{2\pi inu} = \sum a_n e^{-2\pi i\alpha} e^{2\pi i(n-1)u};$$

in other words,

$$a_n e^{2\pi i(n\tau + \alpha)} = a_{n+1}.$$

So if we begin with

$$a_0 = 1.$$

and pick $\alpha = (\tau/2)$, we get

$$a_n = \exp\left\{2\pi i\left(\sum_{k=1}^{n}(k - \tfrac{1}{2})\right)\tau\right\} = \exp\{\pi in^2\tau\},$$

and so

$$f(u) = \sum_{n=-\infty}^{\infty} \exp\{\pi i(n^2\tau + 2nu)\}. \tag{3.12}$$

By (3.10), if u lies in a compact subset of $\mathbb{C}$, then for $|n|$ sufficiently large,

$$|\exp\{\pi i(n^2\tau + 2nu)\}| = \exp\{-n(n\pi \operatorname{Im} \tau + 2 \operatorname{Im} u)\} \leq \exp\{-n(n/2)\pi \operatorname{Im} \tau\}.$$

So the series (3.12) converges absolutely and uniformly on compact subsets, and the holomorphic function (3.12) is denoted by

$$\theta\begin{bmatrix}0\\0\end{bmatrix}(u; \tau) \tag{3.13}$$

or, for the moment, more simply by

$$\theta(u).$$

Notice that $\theta(u)$ is an *even* function.

So from the point of view of Fourier series we have found the closest thing to a holomorphic function on

$$E = \frac{\mathbb{C}}{\mathbb{Z} + \mathbb{Z}\tau}$$

that we can. Suppose we start from another point of view. Recall how we constructed all the meromorphic functions on the *Riemann sphere* $\mathbb{CP}_1$. We can summarize the process in the following steps:

1. Find a homogeneous form (a section of a certain line bundle) on $\mathbb{CP}_1$ with exactly one simple zero, call it X.
2. Operate on this form with the group of automorphisms of $\mathbb{CP}_1$, to get the entire collection of linear forms

$$aX + bY,$$

$a, b \in \mathbb{C}$.

3. Build all meromorphic functions on $\mathbb{CP}_1$ by taking quotients of products

$$f = \frac{(a_1 X + b_1 Y) \cdot \cdots \cdot (a_r X + b_r Y)}{(c_1 X + d_1 Y) \cdot \cdots \cdot (c_r X + d_r Y)}.$$

We will do the analogous process on E, and our function $\theta(u)$ will take the place of the homogeneous form X.

We begin by considering the domain in $\mathbb{C}$, shown in Figure 3.3. This is called a fundamental domain for E since under the mapping (3.5), this domain exactly covers E once except that opposite edges of Figure 3.3 are identified on E. To see how many zeros θ has in Figure 3.3, we compute the boundary integral

$$\frac{1}{2\pi i} \int_{\partial(\text{Figure 3.3})} d \log \theta = \frac{1}{2\pi i} \int_{\alpha+\tau+1}^{\alpha+\tau} (-2\pi i)\, du = 1,$$

since

$$\theta(u + 1) = \theta(u), \qquad \theta(u + \tau) = e^{-\pi i(\tau + 2u)}\theta(u). \qquad (3.14)$$

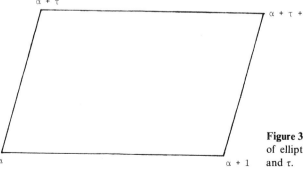

Figure 3.3. Fundamental domain of elliptic curve with periods 1 and τ.

To locate this single zero, we compute

$$\frac{1}{2\pi i} \int_{\partial(\text{Figure 3.3})} u \, d \log \theta$$

$$= \frac{1}{2\pi i} \left[\int_{\alpha+1}^{\alpha+\tau+1} d \log \theta + \int_{\alpha+\tau+1}^{\alpha+\tau} [\tau \, d \log \theta + u(-2\pi i) \, du + \tau(-2\pi i) \, du] \right]$$

$$= \frac{1}{2\pi i} \{ (-\pi i)[(\tau + 2\alpha + 2) + (2 \times \text{integer})] + [\text{integer} \times (2\pi i \tau)]$$

$$+ (\pi i)(2\alpha + 2\tau + 1) + 2\pi i \tau \}$$

$$= (1/2 + \tau/2) + m + n\tau, \qquad m, n \in \mathbb{Z}. \tag{3.15}$$

This places the zero of θ in the middle of the fundamental domain, as shown in Figure 3.4.

Now suppose that

$$p_1, \ldots, p_r \quad \text{and} \quad q_1, \ldots, q_r$$

are points of E such that

$$\sum p_j = \sum q_j,$$

that is, as points of $\mathbb{C}$,

$$\sum p_j - \sum q_j = m + n\tau$$

with $m, n \in \mathbb{Z}$. We then form the function

$$f(u) = \frac{\prod_j \theta(u - p_j - (\frac{1}{2} + \tau/2))}{\prod_j \theta(u - q_j - (\frac{1}{2} + \tau/2))}.$$

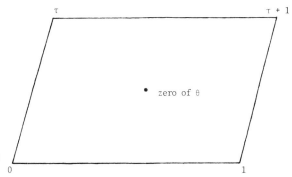

Figure 3.4. The zero of $\theta(u)$ in the fundamental domain.

Clearly $f(u + 1) = f(u)$, but also by (3.14)

$$f(u + \tau)/f(u) = \exp\{2\pi i(\Sigma p_j - \Sigma q_j)\} = \exp\{2\pi i n\tau\}.$$

This means that

$$g(u) = e^{-2\pi i n u}f(u)$$

is *doubly periodic*, that is,

$$g(u + 1) = g(u), \qquad g(u + \tau) = g(u).$$

Thus g gives a well-defined meromorphic function on E with zero set

$$p_1, \ldots, p_r$$

and polar set

$$q_1, \ldots, q_r.$$

Conversely, if $g(u)$ is a meromorphic function on E, with zero set

$$p_1, \ldots, p_r$$

and polar set

$$q_1, \ldots, q_r,$$

then the same computation as (3.15) gives

$$\frac{1}{2\pi i} \int_{\partial(\text{Figure 3.3})} u \, d \log g = m + n\tau$$

for some $m, n \in \mathbb{Z}$. We have therefore proved

> *Abel's theorem for elliptic curves:* There exists a meromorphic function with zero set $\{p_1, \ldots, p_r\}$ and polar set $\{q_1, \ldots, q_r\}$ if and only if
> $$\sum p_j - \sum q_j \in (\mathbb{Z} + \mathbb{Z}\tau).$$

3.4 Meromorphic Functions on Plane Cubics

The meromorphic functions on $\mathbb{CP}_1$ can all be written in the form

$$\frac{p(x, y)}{q(x, y)},$$

where p and q are homogeneous of the same degree. Meromorphic functions on a nonsingular cubic curve $E \subseteq \mathbb{CP}_2$ certainly include the restrictions to E of quotients

$$\frac{p(x, y, z)}{q(x, y, z)}, \tag{3.16}$$

where p and q are again homogeneous of the same degree. By a theorem of Chow (Gunning and Rossi [2], p. 170), all meromorphic functions on $\mathbb{CP}_2$ have the form (3.16), and using the Kodaira vanishing theorem (Hirzebruch [4], p. 140), it can be shown that any meromorphic function on E is the restriction of a function (3.16) to E. However, we will show that all meromorphic functions on E come by restriction of functions (3.16) on $\mathbb{CP}_2$ in a more elementary way. Recall that the group structure on E is also given geometrically in $\mathbb{CP}_2$ (see Chapter Two). The equation

$$p_1 + p_2 = q_1 + q_2 \qquad \text{in } E$$

means that the line L through p_1 and p_2 on E meets E again in the same point that the line M through q_1 and q_2 does (Figure 3.5). If $l(x, y, z) = 0$ is the equation of the line L and $m(x, y, z) = 0$ is the equation of the line M, then the function

$$g(x, y, z) = \frac{l(x, y, z)}{m(x, y, z)}$$

restricts to a meromorphic function in E with zero set $\{p_1, p_2\}$ and polar set

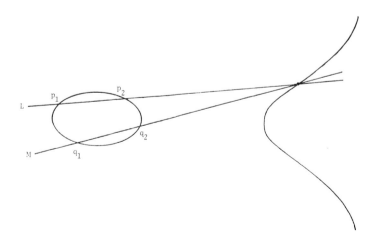

Figure 3.5. The lines L and M have a common point on the cubic.

$\{q_1, q_2\}$. Suppose now that we know that for $r < r_0$, any function with zero set $p_1, \ldots, p_r$ and polar set $q_1, \ldots, q_r$ can be written in the form

$$\frac{f(x, y, z)}{h(x, y, z)},$$

where f and h are homogeneous polynomials of the same degree in (x, y, z). If

$$p_1 + \cdots + p_{r_0} = q_1 + \cdots + q_{r_0} \qquad \text{in } E,$$

then pick p' and q' in E such that

$$p_1 + \cdots + p_{r_0-1} = q_1 + \cdots + q_{r_0-2} + q'$$

and

$$p' + p_{r_0} = q_{r_0-1} + q_{r_0}.$$

Adding these two equations we get

$$p' = q'.$$

Also by the induction hypothesis, we have

$$f_1/h_1$$

with zero set $\{p_1, \ldots, p_{r_0-1}\}$ and polar set $\{q_1, \ldots, q_{r_0-2}, q'\}$, and

$$f_2/h_2$$

with zero set $\{p', p_{r_0}\}$ and polar set $\{q_{r_0-1}, q_{r_0}\}$. Then

$$f_1 f_2/h_1 h_2$$

has zero set $\{p_1, \ldots, p_{r_0}\}$ and polar set $\{q_1, \ldots, q_{r_0}\}$. Thus every meromorphic function on E has the form (3.16).

There is another easy consequence of these theorems. If a meromorphic function f on E has just one simple zero at p and just one simple pole at q, then

$$p = q$$

so that the function has no zeros or poles at all. If f is a meromorphic function on E, we write the formal sum

$$(f) = \sum_{p \in E} (\text{order}_p f)p, \qquad (3.17)$$

which we call the *divisor* of f. Given any $q \in E$, there are many choices of p_1 and p_2 such that

$$p_1 + p_2 = 2q,$$

so there are many meromorphic functions whose only poles are at q and those of order ≤ 2. However, if f and g are any two such, then there are constants a and b such that

$$af + bg$$

has no worse than a first-order pole at q and so is a constant function. Continuing to reason in this way we arrive at the

Riemann–Roch theorem for elliptic curves: If $\sum_{j=1}^{s} r_j \geq 2$ and each $r_j \geq 0$, then the vector space of meromorphic functions with a pole of order no worse than r_j at a point $q_j, j = 1, \ldots, s$, and no other poles has (complex) dimension

$$\sum_{j=1}^{s} r_j.$$

3.5 The Weierstrass p-Function

Suppose we look at a trigonometric function in the following way: The sine function is the inverse function of the function

$$a(\hat{x}) = \int_{x_0}^{\hat{x}} \frac{1}{(1 - x^2)^{1/2}} \, dx.$$

With this in mind it makes sense to hope that the inverse function to

$$b(\hat{x}) = \int_{x_0}^{\hat{x}} \frac{1}{(x^3 + ax^2 + bx + c)^{1/2}} \, dx$$

will also be a reasonably natural function with nice properties. This is indeed the case. When the problem is suitably normalized, its solution is called the Weierstrass p-function, to which we briefly alluded in Chapter Two.

We begin the construction of the p-function with our just-proved Riemann–Roch theorem. The Riemann–Roch theorem implies existence of a meromorphic function on E with 0 its only pole and that of order 2. Let's explicitly construct such a function. First define

$$\theta[{}^1_1](u; \tau) = \sum_{n=-\infty}^{\infty} \exp\{\pi i[(n + \tfrac{1}{2})^2\tau + 2(n + \tfrac{1}{2})(u + \tfrac{1}{2})]\}.$$

This series converges for the same reason that the one defining $\theta[{}^0_0](u; \tau)$ does. Also

$$\theta[{}^1_1](u + 1; \tau) = -\theta[{}^1_1](u; \tau). \tag{3.18}$$

The function $\theta[\begin{smallmatrix}1\\1\end{smallmatrix}]$ is an odd function and so has a zero at 0; and

$$\theta[\begin{smallmatrix}1\\1\end{smallmatrix}](u + \tau; \tau) = (-1)\exp\{-\pi i(\tau + 2u)\}\theta[\begin{smallmatrix}1\\1\end{smallmatrix}](u; \tau) \qquad (3.19)$$

so that just as in (3.14)–(3.16), $\theta[\begin{smallmatrix}1\\1\end{smallmatrix}](u; \tau)$ has its only zeros at the points of $(\mathbb{Z} + \tau\mathbb{Z})$ and a simple zero at each of these. Now

$$\frac{d^2 \log \theta[\begin{smallmatrix}1\\1\end{smallmatrix}](u; \tau)}{du^2}$$

is doubly periodic by (3.18) and (3.19) and has a double pole at 0 since $\theta[\begin{smallmatrix}1\\1\end{smallmatrix}]$ has a simple zero there.

We next investigate periodicity properties of $\theta[\begin{smallmatrix}1\\1\end{smallmatrix}](u; \tau)$ with respect to τ. First we have directly from the definition of $\theta[\begin{smallmatrix}1\\1\end{smallmatrix}](u; \tau)$ that

$$\theta[\begin{smallmatrix}1\\1\end{smallmatrix}](u; \tau + 1) = e^{\pi i/4}\theta[\begin{smallmatrix}1\\1\end{smallmatrix}](u; \tau),$$

and the functions

$$\theta[\begin{smallmatrix}1\\1\end{smallmatrix}](\tau u; \tau) \quad \text{and} \quad \theta[\begin{smallmatrix}1\\1\end{smallmatrix}](u; -1/\tau)$$

both have $(\mathbb{Z} + (-1/\tau)\mathbb{Z})$ as their zero set, both are odd, and both are such that if we apply the operator

$$-\frac{d^2 \log}{du^2}$$

to them, we get doubly-periodic functions whose Laurent series at zero is of the form

$$1/u^2 + (constant) + (even \text{ positive powers of } u).$$

Thus by the Riemann–Roch theorem

$$\frac{d^2 \log \theta[\begin{smallmatrix}1\\1\end{smallmatrix}](\tau u; \tau)}{du^2} = \frac{d^2 \log \theta[\begin{smallmatrix}1\\1\end{smallmatrix}](u; -1/\tau)}{du^2} + c(\tau). \qquad (3.20)$$

If we now develop the Laurent series expansion

$$-\frac{d^2 \log \theta[\begin{smallmatrix}1\\1\end{smallmatrix}](u; \tau)}{du^2} = \frac{1}{u^2} + c_0(\tau) + c_2(\tau)u^2 + \cdots, \qquad (3.21)$$

then (3.21) implies immediately that if $n > 0$,

$$c_{2n}(\tau + 1) = c_{2n}(\tau),$$

$$c_{2n}(-1/\tau) = \tau^{2(n+1)}c_{2n}(\tau). \qquad (3.22)$$

Finally notice that

$$\lim_{\tau \to i \cdot \infty} e^{-\pi i \tau / 4} \theta[{}^1_1](u; \tau) = \exp\{\pi i (u + \tfrac{1}{2})\} + \exp\{-\pi i (u + \tfrac{1}{2})\} = -2 \sin(\pi u)$$

so that

$$\lim_{\tau \to i \cdot \infty} (3.21) = \pi^2 \csc^2(\pi u).$$

Thus all the $c_{2n}(\tau)$ stay bounded as τ goes to infinity along the positive imaginary axis. This last fact, coupled with the relations (3.22), says that the functions $c_{2n}(\tau)$ are *modular forms* (see Serre [8], pp. 77–79). More precisely:

c_{2n} *is a modular form of weight* $(n + 1)$ *[or of weight* $2(n + 1)$ *depending on the convention followed in assigning weights to modular forms].*

Now a very simple residue computation (given in Serre's book [8]), says that the vector space of modular forms of weight 1 has dimension 0 and the vector spaces for weights 2 and 3 each have dimension 1 and their generators together generate the graded ring of modular forms. Thus, up to a constant

$$c_2(\tau) = \sum \{1/(m + n\tau)^4 : (m, n) \in ((\mathbb{Z} \times \mathbb{Z}) - \{(0, 0)\})\},$$
$$c_4(\tau) = \sum 1/(m + n\tau)^6, \tag{3.23}$$

since the convergent series on the right-hand side, called the *Eisenstein series*, are clearly modular forms of the appropriate weights. We define the *Weierstrass p-function*

$$p(u) = -\frac{d^2 \log \theta[{}^1_1](u; \tau)}{du^2} - c_0(\tau)$$

so as to eliminate the troublesome constant term in (3.21).

To see why this function solves the problem posed at the beginning of this section, we again begin with the Riemann–Roch theorem. Namely, if we define the functions

$$1, p, p', p^2, pp', p^3, (p')^2$$

where $p' = dp/du$, we have seven functions on

$$\frac{\mathbb{C}}{\mathbb{Z} + \tau \mathbb{Z}},$$

each of which has no pole other than 0 and with poles of order

$$0, 2, 3, 4, 5, 6, 6$$

at 0. By the Riemann–Roch theorem these functions all lie in a six-dimensional vector space, so there is a relation

$$A + Bp + Cp' + Dp^2 + Epp' + Fp^3 + G(p')^2 = 0.$$

Develop the Laurent series of the left-hand side around $u = 0$:

$$p(u) = \frac{1}{u^2} + c_2 u^2 + c_4 u^4 + \cdots,$$

$$p'(u) = \frac{-2}{u^3} + 2c_2 u + 4c_4 u^3 + \cdots,$$

$$p^2(u) = \frac{1}{u^4} + 2c_2 + 2c_4 u^2 + \cdots,$$

$$p(u)p'(u) = \frac{-2}{u^5} + au + \cdots,$$

$$p^3(u) = \frac{1}{u^6} + \frac{3c_2}{u^2} + 3c_4 + \cdots,$$

$$[p'(u)]^2 = \frac{4}{u^6} - \frac{8c_2}{u^2} - 16c_4 + \cdots.$$

Thus $C = D = E = 0$ and we can normalize by setting $G = -1$ to obtain

$$(p')^2 = 4p^3 - 20c_2 p - 28c_4. \tag{3.24}$$

From this we see that we have a well-defined mapping

$$f : \frac{\mathbb{C}}{\mathbb{Z} + \tau\mathbb{Z}} \longrightarrow E,$$

$$u \longmapsto (p(u), p'(u)) \tag{3.25}$$

into the cubic curve

$$y^2 = 4x^3 - 20c_2(\tau)x - 28c_4(\tau).$$

If we extend to a mapping

$$\frac{\mathbb{C}}{\mathbb{Z} + \tau\mathbb{Z}} \longrightarrow \mathbb{C}\mathbb{P}_2,$$

$$u \longmapsto (p, p', 1),$$

then the mapping is everywhere defined and sends the point 0 to (0, 1, 0) as it should. This mapping is everywhere of maximal rank, since p' is zero only at the four points of ramification of the 2 to 1 covering

$$\frac{\mathbb{C}}{\mathbb{Z} + \tau\mathbb{Z}} \longrightarrow \mathbb{CP}_1,$$

$$u \longmapsto (p(u), 1)$$

and has only simple zeros at each, so p'' is not zero there. Also the mapping (3.25) is injective since it is of maximal rank and proper, therefore a covering space, but $(0, 1, 0)$ has only one preimage. Thus:

Every manifold of the form $C/(\mathbb{Z} + \tau\mathbb{Z})$ is isomorphic to a cubic curve.

If we now look at the differential

$$\frac{dx}{y}$$

on E and pull it back under the map (3.25), we obtain the differential

$$du$$

on the domain. So

$$\int_{x(p_0)}^{x(p)} \frac{dx}{y} = u\left(f^{-1}(p)\right).$$

So (3.25) is the inverse map to

$$q \longmapsto \int_{x(q_0)}^{x(q)} \frac{dx}{(4x^3 - 20c_2 x - 28c_4)^{1/2}}.$$

3.6 Theta-Null Values Give Moduli of Elliptic Curves

The relation (3.20) allows us to conclude that

$$\theta\begin{bmatrix}1\\1\end{bmatrix}(u; -1/\tau) = \alpha(\tau)\exp\{\beta(\tau)u^2 + \gamma(\tau)u\}\theta\begin{bmatrix}1\\1\end{bmatrix}(u\tau; \tau).$$

The famous *Poisson summation formula* from the theory of Fourier series will later allow us to compute

$$\alpha(\tau), \beta(\tau), \gamma(\tau)$$

explicitly. This is an interesting computation which we shall do at the end of this chapter.

But for now notice that the functions

$$\left(\theta\begin{bmatrix}1\\1\end{bmatrix}(u; \tau)\right)^2 \quad \text{and} \quad \left(\theta\begin{bmatrix}0\\0\end{bmatrix}(u; \tau)\right)^2$$

satisfy the same periodicity relations, namely,

$$f(u + 1) = f(u),$$
$$f(u + \tau) = \exp\{-2\pi i(\tau + 2u)\} f(u). \tag{3.26}$$

Again the first relation means that f has a Fourier series expansion and the second relation means that the nth Fourier coefficient determines the $(n + 2)$nd. So the vector space of functions satisfying (3.26) has dimension at most two. Since $(\theta[^1_1](u; \tau))^2$ and $(\theta[^0_0](u; \tau))^2$ are linearly independent, they are a basis of the holomorphic functions on $\mathbb{C}$ which satisfy (3.26). Also we obtain a map

$$\begin{aligned} h: \quad E &\longrightarrow \mathbb{CP}_1, \\ u &\longmapsto (\theta[^0_0](u; \tau)^2, \theta[^1_1](u; \tau)^2). \end{aligned} \tag{3.27}$$

Now $h(\tfrac{1}{2} + \tau/2) = (0, 1)$ by (3.15) and h is simply branched there, as well as at $h(0) = (1, 0)$. Since $h^{-1}((1, 0)) = \{0\}$, we conclude that h is a double cover and is branched over two other points on $\mathbb{CP}_1$. We want to compute the projective coordinates of these two other points. We begin by noticing that

$$\begin{aligned} \mathbb{CP}_1 &\longrightarrow E, \\ q &\longmapsto \sum p, \qquad p \in h^{-1}(q), \end{aligned}$$

lifts to an entire function on $\mathbb{CP}_1$ and is therefore constant. We see that the two other ramification points, p_1 and $p_2 \in E$, at which h is ramified, must have the property that

$$2p_j = 2 \cdot 0 = 2 \cdot (1/2 + \tau/2) = 0.$$

Thus the values of h at these points are

$$(\theta[^0_0](1/2; \tau)^2, \theta[^1_1](1/2; \tau)^2),$$
$$(\theta[^0_0](\tau/2; \tau)^2, \theta[^1_1](\tau/2; \tau)^2).$$

However, for reasons that appear later, it is preferable to have the values (3.27) of the branch locus of h in terms of the *theta-null values*, that is, the four numbers

$$\theta[^0_0](0; \tau), \theta[^0_1](0; \tau), \theta[^1_0](0; \tau), \theta[^1_1](0; \tau)$$

(two of which we have yet to define).

To achieve this, we begin by defining the even function

$$\theta[^0_1](u; \tau) = \theta[^0_0](u + \tfrac{1}{2}; \tau) = \sum \exp\{\pi i[n^2\tau + 2n(u + \tfrac{1}{2})]\},$$

which satisfies the transformation laws

$$\theta[{}^0_1](u + 1; \tau) = \theta[{}^0_1](u, \tau),$$
$$\theta[{}^0_1](u + \tau; \tau) = -\exp\{-\pi i(\tau + 2u)\}\theta[{}^0_1](u; \tau),$$

and the even function

$$\theta[{}^1_0](u; \tau) = \theta[{}^1_1](u - \tfrac{1}{2}; \tau) = \sum \exp\{\pi i[(n + \tfrac{1}{2})^2\tau + 2(n + \tfrac{1}{2})u]\},$$

which satisfies the transformation laws

$$\theta[{}^1_0](u + 1; \tau) = -\theta[{}^1_0](u; \tau),$$
$$\theta[{}^1_0](u + \tau; \tau) = \exp\{-\pi i(\tau + 2u)\}\theta[{}^1_0](u; \tau).$$

That is, for any δ, $\varepsilon \in \{0, 1\}$,

$$\theta[{}^\delta_\varepsilon](u + 1; \tau) = (-1)^\delta\theta[{}^\delta_\varepsilon](u; \tau),$$
$$\theta[{}^\delta_\varepsilon](u + \tau; \tau) = (-1)^\varepsilon \exp\{-\pi i(\tau + 2u)\}\theta[{}^\delta_\varepsilon](u; \tau).$$

Also by direct computation it is easy to see that

$$\theta[{}^0_0](\tau/2; \tau) = \exp\{-\pi i\tau/4\}\theta[{}^1_0](0; \tau),$$
$$\theta[{}^1_1](\tau/2; \tau) = \exp\{-\pi i(\tau + 2)/4\}\theta[{}^0_1](0; \tau).$$

So now we have the other two branch points of the map h in (3.27) given in terms of theta-nulls. Namely, they are the points

$$(\theta[{}^0_1]^2, \theta[{}^1_0]^2),$$
$$(-\theta[{}^1_0]^2, \theta[{}^0_1]^2),$$

where

$$\theta[{}^\delta_\varepsilon] = \theta[{}^\delta_\varepsilon](0; \tau).$$

Thus our curve E is isomorphic to the curve

$$y^2 = x(x - 1)(x - \lambda),$$

where

$$\lambda = -\theta[{}^1_0]^4/\theta[{}^0_1]^4. \qquad (3.28)$$

In fact, then, all six cross-ratios

$$\lambda, \frac{1}{\lambda}, 1 - \lambda, \frac{1}{1 - \lambda}, \frac{\lambda}{\lambda - 1}, \frac{\lambda - 1}{\lambda}$$

which uniquely determine E are given in terms of theta-null values.

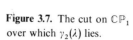

Figure 3.6. The cut on $\mathbb{CP}_1$
over which $\gamma_1(\lambda)$ lies.

Figure 3.7. The cut on $\mathbb{CP}_1$
over which $\gamma_2(\lambda)$ lies.

3.7 The Moduli Space of "Level-Two Structures" on Elliptic Curves

Suppose that we denote the curve

$$y^2 = x(x - 1)(x - \lambda),$$

or rather its closure in $\mathbb{CP}_2$, by $E(\lambda)$. In Chapter Two we integrated the differential form dx/y over the basis

$$\gamma_1(\lambda) \quad \text{and} \quad \gamma_2(\lambda) \tag{3.29}$$

of $H_1(E(\lambda); \mathbb{Z})$. The cycles (3.29) were given by the inverse images in $E(\lambda)$ of the slits on $\mathbb{CP}_1$ shown in Figures 3.6 and 3.7. As λ moves, so do the one-cycles (3.29). Clearly we can move λ around, then return it to its original position, to achieve the homology transformations

$$\gamma_1(\lambda) \longmapsto \gamma_1(\lambda),$$
$$\gamma_2(\lambda) \longmapsto \gamma_2(\lambda) + 2\gamma_1(\lambda)$$

and

$$\gamma_1(\lambda) \longmapsto \gamma_1(\lambda) + 2\gamma_2(\lambda),$$
$$\gamma_2(\lambda) \longmapsto \gamma_2(\lambda).$$

The first transformation is achieved by moving λ *around* the slit $\overline{01}$ once and back to its starting position, the second by moving λ *through* that slit and back to its original position. These two transformations, given by the matrices

$$\begin{bmatrix} 1 & 2 \\ 0 & 1 \end{bmatrix} \quad \text{and} \quad \begin{bmatrix} 1 & 0 \\ 2 & 1 \end{bmatrix},$$

generate the group of transformations of the λ plane given by

$$\Gamma_2 = \left\{ \begin{bmatrix} a & b \\ c & d \end{bmatrix} \in \mathrm{Sp}_1(\mathbb{Z}): \begin{bmatrix} a & b \\ c & d \end{bmatrix} \equiv \begin{bmatrix} 1 & 0 \\ 0 & 1 \end{bmatrix} \mod 2 \right\},$$

integer-valued matrices where

$$\mathrm{Sp}_1(\mathbb{Z}) = \left\{ \begin{bmatrix} a & b \\ c & d \end{bmatrix}: ad - bc = 1 \right\}.$$

To see this, multiply $\begin{bmatrix} a & b \\ c & d \end{bmatrix}$ on the right by $\begin{bmatrix} 1 & 0 \\ 2r & 1 \end{bmatrix}$ to get $|a| < |b|$, by $\begin{bmatrix} 1 & 2s \\ 0 & 1 \end{bmatrix}$ to get $|b| < |a|$, etc., until $b = 0$. On the other hand, it is clear that no matter how we move λ around and return it to its original position, the homology transformation will be the identity modulo 2. We say that the *monodromy group* of the family $\{E(\lambda)\}$ is Γ_2.

We have seen that

$$E(\lambda) \cong E\left(\frac{1}{\lambda}\right) \cong E(1 - \lambda) \cong E\left(\frac{1}{1 - \lambda}\right) \cong E\left(\frac{\lambda}{\lambda - 1}\right) \cong E\left(\frac{\lambda - 1}{\lambda}\right).$$

Thus λ does not exactly measure the isomorphism class of the curve

$$y^2 = x(x - 1)(x - \lambda).$$

What then does λ measure? If we are given λ, then we are given not only $E(\lambda)$ but also a specific choice of the identity element of the group $E(\lambda)$ (namely, the point at infinity), as well as a *specific ordering* of the points of order 2 on $E(\lambda)$, namely

$$(0, 0)\ (1, 0)\ (\lambda, 0). \tag{3.30}$$

Now to get the cross ratio

$$\frac{z_3 - z_1}{z_3 - z_2} : \frac{z_4 - z_1}{z_4 - z_2}$$

to have the value λ, we order our branch points as follows:

$$z_1 = \infty, \qquad z_2 = 0, \qquad z_3 = 1, \qquad z_4 = \lambda.$$

To get $1/\lambda$ we use the ordering

$$z_1 = \infty, \qquad z_2 = 0, \qquad z_3 = \lambda, \qquad z_4 = 1,$$

and to get $(1 - \lambda)$ we use

$$z_1 = \infty, \qquad z_2 = 1, \qquad z_3 = 0, \qquad z_4 = \lambda.$$

So under the isomorphism

$$E(\lambda) \cong E(1/\lambda)$$

the first point of order 2 goes to the first, but the second and third are interchanged. Similarly, under

$$E(\lambda) \cong E(1 - \lambda),$$

the first two points of order 2 are interchanged. Since the isomorphisms to

$$E\left(\frac{1}{1 - \lambda}\right),\ E\left(\frac{\lambda}{\lambda - 1}\right),\ E\left(\frac{\lambda - 1}{\lambda}\right)$$

are built up by composition, we conclude that the six distinct cross ratios corresponding to the same elliptic curve correspond to the six distinct orderings of the points of order 2, or, what is the same thing, to the six distinct choices of ordered bases for $H_1(E(\lambda); \mathbb{F}_2)$.

Let's call a choice of ordered basis of $H_1(E(\lambda); \mathbb{F}_2)$ a *framing*. Then we say that λ gives us the curve $E(\lambda)$ together with a framing of $H_1(E(\lambda); \mathbb{F}_2)$. Conversely, given a cubic curve E and a framing of $H_1(E; \mathbb{F}_2)$, there is only one choice of λ such that the framing determined by λ is the given one. But we must be a little careful, for there are two cases in which the six numbers

$$\lambda, \frac{1}{\lambda}, 1 - \lambda, \frac{1}{1 - \lambda}, \frac{\lambda}{\lambda - 1}, \frac{\lambda - 1}{\lambda}$$

are not all distinct among themselves. First, we have the case in which the six numbers are

$$-1, -1, 2, \tfrac{1}{2}, \tfrac{1}{2}, 2,$$

that is, the case of the curve

$$y^2 = x(x^2 - 1). \tag{3.31}$$

Second, we have the case in which the set of six numbers is

$$\rho, \ \bar{\rho}, \ \bar{\rho}, \ \rho, \ \bar{\rho}, \ \rho,$$

where ρ is a primitive cube root of -1. This set of cross ratios corresponds to the curve

$$y^2 = x^3 - 1. \tag{3.32}$$

Now the curve (3.31) clearly admits the automorphism

$$(x, y) \mapsto (-x, iy), \tag{3.33}$$

which does *not* induce the identity map on $H_1(E; \mathbb{F}_2)$. This automorphism takes a framing of $H_1(E; \mathbb{F}_2)$ into a distinct one. Since the automorphism (3.33) is induced by lifting a linear fractional transformation on $\mathbb{CP}_1$, the cross ratio must be preserved. Similarly, the curve (3.32) admits the automorphisms

$$(x, y) \mapsto (\sigma x, -y),$$

where σ is either primitive cube root of unity. These automorphisms also act nontrivially on all framings and preserve cross ratios.

Now suppose that we have *any* isomorphism

$$\alpha \colon E(\lambda_0) \cong E(\lambda_0)$$

which takes the identity 0 in E to 0. We then have the diagram

$$
\begin{array}{ccc}
E(\lambda_0) & \xrightarrow{\ \alpha\ } & E(\lambda_0) \\
\downarrow (f,\,1) & & \downarrow (f,\,1) \\
\mathbb{CP}_1 & & \mathbb{CP}_1
\end{array}
$$

where f is a two-sheeted covering branched over ∞ with $f(0) = \infty$. By the Riemann–Roch theorem,

$$
f \cdot \alpha = cf + d
$$

for some constants c, d, and so we can complete the preceding diagram to obtain

$$
\begin{array}{ccc}
E(\lambda_0) & \xrightarrow{\ \alpha\ } & E(\lambda_0) \\
\downarrow f & & \downarrow f \\
\mathbb{CP}_1 & \xrightarrow{\ \beta\ } & \mathbb{CP}_1
\end{array}
$$

where β is a linear fractional transformation. If α changes the framing of $H_1(E(\lambda_0); \mathbb{F}_2)$, then since β preserves cross ratios, the set of six cross-ratios must have repeated numbers. So $E(\lambda_0)$ must be one of the two curves (3.31) or (3.32). We can sum up this discussion as follows:

There is a one-to-one correspondence between the set of complex analy-tic isomorphism classes of pairs

$$(E, \Gamma) \tag{3.34}$$

[where E is a cubic curve and Γ is a framing of $H_1(E; \mathbb{F}_2)$] and the set

$$\mathbb{C} - \{0, 1\}.$$

3.8 Automorphisms of Elliptic Curves

In fact, we can list all automorphisms of elliptic curves. Let

$$\alpha \colon E \to E$$

be an automorphism of an elliptic curve E. Having coupled α with an appropriate translation, we can assume

$$\alpha(0) = 0.$$

If α acts trivially on $H_1(E; \mathbb{F}_2)$, then in terms of a fundamental domain for

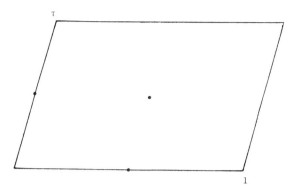

Figure 3.8. Fundamental domain for elliptic curve E with points of order 2 marked.

E (Figure 3.8) the points $1/2$, $\tau/2$, $1/2 + \tau/2$, and 0 must all be fixed in E under the action of α. So α acts on line segments as follows

$$\overline{(0, 1/2)} \mapsto \overline{(0, 1/2)} \quad \text{or} \quad \overline{(1, 1/2)},$$

$$\overline{(0, \tau/2)} \mapsto \overline{(0, \tau/2)} \quad \text{or} \quad \overline{(\tau, \tau/2)}.$$

(Remember that α is linear.) Since α preserves orientation, the only possibilities are

$$\alpha = \pm(\text{identity map}).$$

From what we have done we can make a complete list of the complex analytic automorphisms of cubic curves:

(a) translations
(b) $\pm(\text{identity map})$
(c) the curve $y^2 = x(x^2 - 1)$, which admits an automorphism whose square is $-(\text{identity})$
(d) the curve $y^2 = x^3 - 1$, which admits an automorphism whose cube is $-(\text{identity})$

All automorphisms of a given curve are obtained by composing automorphisms of that curve which appears on this list.

3.9 The Moduli Space of Elliptic Curves

Next suppose we let

$$\mathcal{H} = \{\tau \in \mathbb{C} : \operatorname{Im} \tau > 0\},$$

the *upper half-plane*. We then have a natural map

$$\mathcal{H} \longrightarrow \mathbb{C} - \{0, 1\},$$
$$\tau \longmapsto \left(-\theta[{}^1_0]^4/\theta[{}^0_1]^4\right)$$

(3.35)

[see (3.28)]. We write

$$E_\tau = \frac{\mathbb{C}}{\mathbb{Z} + \tau \mathbb{Z}}.$$

Then we have seen in (3.3) that

$$E_\tau \cong E_{\tau'}$$

if and only if there is a complex constant e such that

$$e(\mathbb{Z} + \mathbb{Z}\tau) = (\mathbb{Z} + \mathbb{Z}\tau').$$

This will happen if and only if the equations

$$e(d + c\tau) = 1,$$
$$e(b + a\tau) = \tau'$$

(3.36)

can be solved, where, as in Section 3.7,

$$\begin{bmatrix} a & b \\ c & d \end{bmatrix} \in \mathrm{Sp}_1(\mathbb{Z}).$$

This happens if and only if

$$\tau' = \frac{a\tau + b}{c\tau + d}.$$

So we define an action of $\mathrm{Sp}_1(\mathbb{Z})$ on $\mathcal{H}$ by the rule

$$\begin{bmatrix} a & b \\ c & d \end{bmatrix} \cdot \tau = (a\tau + b)(c\tau + d)^{-1},$$

(3.37)

and we conclude that the set of orbits of this action is in one-to-one correspondence with the set of isomorphism classes of elliptic curves, that is,

$$\mathcal{M} \underset{(\text{def})}{=} (\text{isomorphism classes of elliptic curves}) \cong \mathrm{Sp}_1(\mathbb{Z}\backslash\mathcal{H}). \quad (3.38)$$

To sum up, we have natural maps

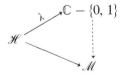

As we have set things up, the solid arrows are holomorphic maps. We next want to study the structure of the mapping given by the broken arrow.

3.10 And So, by the Way, We Get Picard's Theorem

Let

$$\tau : \mathscr{E} \to \mathscr{H} \tag{3.39}$$

be the natural bundle with fibre E_τ for each $\tau \in \mathscr{H}$. We can then define an action of $\mathrm{Sp}_1(\mathbb{Z})$ on $\mathscr{E}$ which is compatible with the action defined by (3.37); namely, we define the isomorphism

$$E_\tau \to E_{\tau'},$$
$$u \mapsto eu, \tag{3.40}$$

where

$$e = 1/(c\tau + d),$$

as in (3.36).

If $\lambda(\tau)$ is as in (3.28) and $\tau \neq \tau'$ but

$$\lambda(\tau) = \lambda(\tau'),$$

then there is an element $A \in \mathrm{Sp}_1(\mathbb{Z})$ such that

$$A \cdot \tau = \tau'.$$

Also we have arranged things so that under the isomorphism

$$E_\tau \to E\big(\lambda(\tau)\big)$$

given by normalizing (3.27),

$$0 \longmapsto \infty,$$
$$1/2 \longmapsto 1,$$
$$1/2 + \tau/2 \longmapsto 0,$$
$$\tau/2 \longmapsto \lambda(\tau).$$

Therefore the map (3.40) must preserve points of order 2. From (3.36)

$$\tfrac{1}{2} \longmapsto (-c\tau' + a)/2 \equiv \tfrac{1}{2}$$
$$\tau/2 \longmapsto (d\tau' - b)/2 \quad \equiv \tau'/2$$

Therefore

$$A \equiv I \bmod 2, \qquad \text{i.e., } A \in \Gamma_2. \tag{3.41}$$

On the other hand, if (3.41) is satisfied, then the composition

$$E(\lambda(\tau)) \cong E_\tau \cong E_{\tau'} \cong E(\lambda(\tau'))$$

preserves the ordering of the points of order 2 so that

$$\lambda(\tau) = \lambda(\tau').$$

Therefore we have an injection

$$\lambda : \Gamma_2 \backslash \mathcal{H} \to \mathbb{C} - \{0, 1\}. \tag{3.42}$$

To see that the mapping (3.42) is actually a surjection, we reason as follows. For any

$$\lambda_0 \in \mathbb{C} - \{0, 1\},$$

there must exist a $\tau \in \mathcal{H}$ such that

$$E_\tau \cong E(\lambda_0).$$

So

$$E(\lambda(\tau)) \cong E(\lambda_0),$$

from which we conclude that $\lambda(\tau)$ is in the set of six cross ratios determined by λ_0. Now the group of order 6,

$$\mathrm{Sp}_1(\mathbb{Z})/\Gamma_2,$$

acts faithfully on the six orderings of the points of order 2, so there must exist

$$A \in \mathrm{Sp}_1(\mathbb{Z})$$

such that if

$$\tau' = A \cdot \tau,$$

then the isomorphism

$$E(\lambda(\tau')) \cong E(\lambda_0)$$

preserves the distinguished ordering of the points of order 2. So

$$\lambda_0 = \lambda(\tau').$$

Next suppose there exists

$$A \in \mathrm{Sp}_1(\mathbb{Z}), \qquad \tau \in \mathcal{H},$$

such that

$$A \cdot \tau = \tau.$$

Then, referring to (3.36), we see that the isomorphism

$$E_\tau \to E_\tau,$$

$$u \to eu$$

is not $\pm$(identity) unless

$$c = b = 0, \qquad a = d = \pm 1.$$

Thus either $A = \pm$(identity) or the curve E_τ has a nontrivial automorphism. In the latter case, we refer to Section 3.8 to conclude that

$$\tau = i \text{ or } \rho.$$

In fact, from Section 3.7 we can conclude more. Every nontrivial automorphism of E_i and E_ρ gives a nontrivial permutation of the points of order 2, so that $A \notin \Gamma_2$. Our final conclusion is therefore that Γ_2 acts on $\mathscr{H}$ without fixed points so that

$$\lambda : \mathscr{H} \to (\mathbb{C} - \{0, 1\})$$

is a covering space, in fact, the realization of the half-plane as the universal covering space of the complex line minus two points.

   An aside: A trivial corollary of this last fact is *Picards' theorem* that any entire function which omits at least two values is constant. This is because the two omitted values can be assumed to be 0 and 1 and, by the theory of covering spaces, the map then lifts to an analytic map to the universal covering space $\mathscr{H}$ of $\mathbb{C} - \{0, 1\}$. But $\mathscr{H}$ is analytically equivalent to the unit disk so the function must have been constant.

3.11 The Complex Structure of $\mathscr{M}$

We saw in Section 3.10 that there is a natural map

$$j : (\mathbb{C} - \{0, 1\}) \longrightarrow \mathscr{M}$$
$$\lambda \longrightarrow \text{(isomorphism class of } E(\lambda)]. \qquad (3.43)$$

It is just the natural quotient map

$$\Gamma_2 \backslash \mathscr{H} \to \mathrm{Sp}_1(\mathbb{Z}) \backslash \mathscr{H}$$

induced by the action of the group

$$\mathrm{Sp}_1(\mathbb{Z}) \backslash \Gamma_2.$$

Since the map is "generically" six-to-one, the quotient space of $(\mathbb{C} - \{0, 1\})$

with respect to the action of this group is a one-dimensional complex manifold $\mathscr{M}$ whose topological Euler characteristic χ satisfies the relation

$$6\chi - 7 = -1.$$

Thus $\chi = 1$ and so the manifold is simply connected. The points of $\mathscr{M}$ are in one-to-one correspondence with the set of isomorphism classes of cubic curves, so it is called a *moduli space*. By the uniformization theorem for Riemann surfaces (Springer, *Introduction to Riemann Surfaces* [9], p. 224), $\mathscr{M}$ is either $\mathbb{C}$ or the unit disk. Since $(\mathbb{C} - \{0, 1\})$ admits no bounded analytic functions,

$$\mathscr{M} \cong \mathbb{C}.$$

In fact, we can see this without using the uniformization theorem. We check directly that the rational function

$$\phi(\lambda) = \frac{(\lambda^2 - \lambda + 1)^3}{\lambda^2(\lambda - 1)^2} \qquad (3.44)$$

is invariant under the substitutions

$$\lambda \to 1/\lambda \quad \text{and} \quad \lambda \to 1 - \lambda$$

and so also under the substitutions which send λ to $1/(1 - \lambda)$, $\lambda/(\lambda - 1)$, and $(\lambda - 1)/\lambda$ respectively. Since ϕ is six-to-one, it follows that $\mathscr{M}$ is isomorphic to the image of the mapping

$$\phi \colon (\mathbb{C} - \{0, 1\}) \to \mathbb{C}.$$

Since ϕ is a nonconstant algebraic function, the image omits at most a finite set, and since the image is simply connected, it must in fact be equal to all of $\mathbb{C}$. Notice also that by formula (3.28) the point in $\mathscr{M}$ corresponding to a cubic curve E is determined by its theta-nulls.

Now if we have an algebraic (analytic) family of cubic curves in $\mathbb{CP}_2$, that is, a family $\{E_t\}$ of cubic curves

$$\sum_{i+j+k=3} \alpha_{ijk}(t) x^i t^j z^k = 0,$$

where the $\alpha_{ijk}(t)$ are algebraic (analytic) functions of some complex parameter t, then by the implicit function theorem we have the following:

1. There is an algebraic† (analytic) function

$$\beta \colon (t\text{-line}) \to \mathbb{CP}_2$$

---

† This function is an "implicit" and therefore possibly multivalued algebraic function.

such that $\beta(t)$ is always one of the nine inflection points of E_t (see Chapter Two).

2. Using $\beta(t)$ as center of projection, one sees that the function

$$t \mapsto \lambda(t) = \text{cross ratio of branch points of } E_t$$

is an algebraic (analytic) function, since the branch-point set is given by the zero set of a discriminant polynomial whose coefficients are algebraic (analytic) functions of t.

3. The induced map

$$(t\text{-line}) \to \mathcal{M}$$

is algebraic (analytic).

It is the last property which makes $\mathcal{M}$ a "good" moduli space, and this property characterizes $\mathcal{M}$ uniquely. Notice also that if we have an analytic family

$$\frac{\mathbb{C}}{\mathbb{Z} + \tau(t)\mathbb{Z}},$$

where $\tau(t)$ is an analytic function of t, then the formulas (3.28) and (3.44) imply that the induced mapping of the t line into $\mathcal{M}$ is analytic.

3.12 The j-Invariant of an Elliptic Curve

We now return to a topic touched on in Section 3.5 and earlier. If E is a cubic or elliptic, curve, then we chose a *symplectic* basis for $H_1(E; \mathbb{Z})$, namely, a pair of cycles γ_1 and γ_2 such that

$$\gamma_1 \cdot \gamma_2 = +1.$$

Given another choice, δ_1 and δ_2, then

$$\delta_1 = a\gamma_1 - c\gamma_2,$$
$$\delta_2 = (-b)\gamma_1 + d\gamma_2,$$

where a, b, c, and $d \in \mathbb{Z}$ and

$$\delta_1 \cdot \delta_2 = (ad - bc) = +1.$$

The group of such matrices $\begin{bmatrix} a & b \\ c & d \end{bmatrix}$ is called

$$\text{Sp}_1(\mathbb{Z}) \qquad \text{or} \qquad \text{SL}_2(\mathbb{Z}).$$

It is the automorphism group of

$$(H_1(E; \mathbb{Z}), \text{intersection pairing}).$$

Now we obtained the number τ in forming the model

$$\frac{\mathbb{C}}{\mathbb{Z} + \tau\mathbb{Z}}$$

for E as

$$\tau = \int_{\gamma_2} \omega \bigg/ \int_{\gamma_1} \omega, \qquad (3.45)$$

where ω is the everywhere-holomorphic differential on E. Since ω has no zeros, it is unique up to a multiplicative constant by the maximum principle. Thus τ is completely determined by E itself modulo the action

$$\tau \mapsto \frac{b \int_{\gamma_1} \omega + a \int_{\gamma_2} \omega}{d \int_{\gamma_1} \omega + c \int_{\gamma_2} \omega} = \frac{a\tau + b}{c\tau + d}$$

of $\mathrm{Sp}_1(\mathbb{Z})$. Also each number in the upper half-plane occurs in a formula (3.46), since given τ we can take

$$E = \frac{\mathbb{C}}{\mathbb{Z} + \tau\mathbb{Z}},$$

$\omega = du$, the standard differential on $\mathbb{C}$, and γ_1, γ_2 the edges of the fundamental domain. Thus there is a one-to-one correspondence between $\mathcal{M}$, the set of isomorphism classes of elliptic curves, and the set

$$\mathrm{Sp}_1(\mathbb{Z}) \backslash \mathcal{H} \qquad (3.46)$$

where $\mathcal{H}$ is the upper half-plane.

In Serre's book [8], pp. 77–79, it is shown that $\mathrm{Sp}_1(\mathbb{Z})$ acts properly discontinuously with fundamental domain shown in Figure 3.9. It is also shown that

$$\mathrm{Sp}_1(\mathbb{Z})/\{\pm \text{identity}\}$$

is generated by the elements

$$\begin{bmatrix} 0 & 1 \\ -1 & 0 \end{bmatrix} \quad \text{and} \quad \begin{bmatrix} 1 & 1 \\ 0 & 1 \end{bmatrix},$$

that is, by the transformations

$$\begin{aligned} \tau &\mapsto -1/\tau, \\ \tau &\mapsto \tau + 1. \end{aligned} \qquad (3.47)$$

Using this information we obtain another way of embedding

$$\mathrm{Sp}_1(\mathbb{Z}) \backslash \mathcal{H} = \mathcal{M}$$

in projective space. Recall that we found, in (3.22), functions

$$c_{2n}: \mathcal{H} \to \mathbb{C}$$

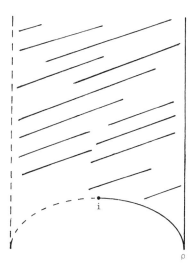

Figure 3.9. Fundamental domain for the action of $Sp_1(\mathbb{Z})$ on the upper half-plane.

such that

$$c_{2n}(-1/\tau) = \tau^{2(n+1)}c_{2n}(\tau),$$
$$c_{2n}(\tau + 1) = c_{2n}(\tau).$$

And so, since the two transformations (3.47) generate the action of $Sp_1(\mathbb{Z})$ on $\mathcal{H}$.

$$c_{2n}\left(\frac{a\tau + b}{c\tau + d}\right) = (c\tau + d)^{2(n+1)}c_{2n}(\tau)$$

for all $\begin{bmatrix} a & b \\ c & d \end{bmatrix} \in Sp_1(\mathbb{Z})$. Since the c_{2n} are also bounded as $\tau \to i \cdot \infty$, we say that c_{2n} is an *automorphic form* of weight $(n + 1)$. Clearly the set of automorphic forms generates a graded algebra, and if

$$f_0, \ldots, f_n$$

are all homogeneous elements of this algebra of the same weight, then

$$\mathcal{H} \longrightarrow \mathbb{CP}_n,$$
$$\tau \longmapsto (f_0(\tau), \ldots, f_n(\tau))$$

descends to a well-defined mapping

$$Sp_1(\mathbb{Z})\backslash\mathcal{H} \to \mathbb{CP}_n.$$

Actually, the easy residue computation given in Serre's book [8] shows that if

$$M_k \tag{3.48}$$

is the vector space of modular forms of weight k and f is a nonzero element of M_k, then

$$\text{ord}_\infty f + \tfrac{1}{2} \text{ord}_i f + \tfrac{1}{3} \text{ord}_\rho f$$
$$+ [\text{sum of orders of } f \text{ at other points of } \mathscr{H}/\text{Sp}_1(\mathbb{Z})] = k/6.$$

(Use the uniformizing parameter $q = e^{2\pi i\tau}$ to compute ord_∞.) Thus c_2 has its only zero at ρ and has a simple zero there, c_4 has its only zero at i and has a simple zero there, and some linear combination

$$\Delta(\tau) = Ac_2(\tau)^3 + Bc_4(\tau)^2 \tag{3.49}$$

is the unique modular form (up to multiplicative constant) of weight 6 with a zero at infinity, and that zero is simple. Now let us form the mapping

$$\text{Sp}_1(\mathbb{Z})\backslash\mathscr{H} \longrightarrow \mathbb{CP}_1.$$
$$\tau \longmapsto (c_2(\tau)^3, \Delta(\tau)). \tag{3.50}$$

So:

1. If we follow Serre's book [8] and add the point ∞ to $\text{Sp}_1(\mathbb{Z})\backslash\mathscr{H}$ by the rule that ∞ is given by $q = 0$ in the q-disk and

$$q = e^{2\pi i\tau}.$$

   then we obtain a compact complex manifold of dimension 1.
2. The mapping (3.50) extends to an everywhere-defined morphism

$$j: \mathscr{M} \to \mathbb{CP}_1$$

such that the point $(c_2(\infty)^3, 0)$ has only one inverse image.

Thus j must be an isomorphism. So we see in another way that

$$(\text{Sp}_1(\mathbb{Z})\backslash\mathscr{H}) \cong \mathbb{C}$$

via the mapping

$$\tau \mapsto c_2^3(\tau)/\Delta(\tau).$$

This last function of τ is called the *j-invariant* of the elliptic curve associated with τ and is the complete invariant of that curve.

To compute values for A and B in (3.49), recall that just after (3.22) we computed that $c_2(\infty)/\pi^2$ and $c_4(\infty)/\pi^2$ are given by the second- and fourth-order terms, respectively, of the Laurent-series expansion for $\csc^2(\pi u)$. Thus

$$c_2(\infty) = \pi^4/15,$$

$$c_4(\infty) = 2\pi^6/189. \tag{3.51}$$

So we can use

$$\Delta(\tau) = (2^2 \cdot 5^3)c_2(\tau)^3 - (3^3 \cdot 7^2)c_4(\tau)^2.$$

Notice that $\Delta(\tau)$ is nothing more than the discriminant of the cubic polynomial

$$4x^3 - 20c_2 x - 28c_4.$$

Since we saw in (3.24) that the elliptic curve corresponding to the period τ is simply the curve C given by

$$y^2 = 4x^3 - 20c_2 x - 28c_4,$$

we see that if $\tau \to i \cdot \infty$, then two of the four branch points (of C over the x axis) come together. In fact, it is easy to see directly that if, say, $\lambda \to 0$, then

$$\int_1^\lambda \frac{dx}{[x(x-\lambda)(x-1)]^{1/2}}$$

becomes infinite as it should.

3.13 Theta-Nulls as Modular Forms

We will end this chapter with a few more illustrative relations between the coefficients of theta functions which again show their connection with the theory of modular forms. To see to what extent the coefficients of theta functions are modular forms, we must compare

$$\theta[^\delta_\varepsilon](u; \tau) \quad \text{and} \quad \theta[^\delta_\varepsilon](u; \tau + 1)$$

as well as

$$\theta[^\delta_\varepsilon](u; \tau) \quad \text{and} \quad \theta[^\delta_\varepsilon](u; -1/\tau).$$

Let's just begin computing.

First, the relation (3.20) allows us to conclude that

$$\theta[^1_1](u; -1/\tau) = \alpha_1(\tau)\exp\{\beta(\tau)u^2 + \gamma(\tau)u\}\theta[^1_1](\tau u; \tau).$$

Now since $\theta[^1_1](u; \tau)$ is an odd function in u, we can forget about the term $e^{\gamma(\tau)u}$ and conclude that

$$\theta[^1_1](u; -1/\tau) = \alpha_1(\tau)\exp\{\beta(\tau)u^2\}\theta[^1_1](\tau u; \tau).$$

Now $\beta(\tau)$ will be determined by the fact that both sides of this last equation must have the same multipliers, that is, must behave in the same way under the transformation

$$u \mapsto u + 1,$$

$$u \mapsto u + \tau.$$

So using (3.18) and (3.19) we obtain that

$$1 = \exp\{\beta(\tau)(2u + 1)\}\exp\{-\pi i[\tau(2u + 1)]\}$$

so that

$$\beta(\tau) = \pi i\tau.$$

Thus

$$\theta[{}^1_1](u; \ -1/\tau) = \alpha_1(\tau)\exp\{\pi i\tau u^2\}\theta[{}^1_1](\tau u; \ \tau), \tag{3.52}$$

and we have only to compute $\alpha_1(\tau)$. This turns out to be, in fact, the most interesting part of the computation.

We begin by making the computations for $\theta[{}^0_0](u; \ \tau)$ exactly analogous to those which led to (3.52) in the case of $\theta[{}^1_1](u; \ \tau)$. We obtain

$$\theta[{}^0_0](u; \ -1/\tau) = \alpha_0(\tau)\exp\{\pi i\tau u^2\}\theta[{}^0_0](\tau u; \ \tau).$$

We next compute $\alpha_0(\tau)$. We simplify by setting $u = 0$ to obtain

$$\sum \exp\{\pi i[n^2(-1/\tau)]\} = \alpha_0(\tau) \sum \exp\{\pi i(n^2\tau)\}.$$

Next notice that $\alpha_0(\tau)$ is an analytic function of τ so that it suffices to compute it for values

$$\tau = ix, \qquad x \text{ real} > 0.$$

So we are reduced to the equation

$$\sum \exp\{-\pi n^2/x\} = \alpha_0(ix) \sum \exp\{-\pi n^2 x\}. \tag{3.53}$$

Next let's compute the Fourier transform of the function

$$f(t) = \exp\{-\pi x t^2\}$$

$$\hat{f}(s) = \int_{-\infty}^{\infty} e^{-\pi x t^2} e^{-2\pi i s t} \, dt$$

$$= \exp\{-\pi s^2/x\} \cdot \int_{-\infty}^{\infty} \exp\{-[\pi^{1/2} x^{1/2}(t + is/x)]^2\} \, dt$$

$$= \pi^{-1/2} x^{-1/2} e^{-\pi s^2/x} \int_{-\infty}^{\infty} e^{-t^2} \, dt$$

$$= 2\pi^{-1/2} x^{-1/2} e^{-\pi s^2/x} \left[\int_0^{\pi/2} \int_0^{\infty} e^{-r^2} r \, dr \, d\theta \right]^{1/2}$$

$$= x^{-1/2} e^{-\pi s^2/x}.$$

It is here that we apply the *Poisson summation formula* to conclude that

$$\sum \exp\{-\pi x n^2\} = x^{-1/2} \sum \exp\{-\pi n^2/x\}.\dagger$$

Comparing with (3.53), we obtain that

$$\alpha_0(ix) = x^{1/2}$$

so that

$$\theta[{}^0_0](u; -1/\tau) = (\tau/i)^{1/2} \exp\{\pi i u^2 \tau\}\theta[{}^0_0](\tau u; \tau).$$

Also

$$\begin{aligned}
\theta[{}^0_1](u; -1/\tau) &= \theta[{}^0_0](u + \tfrac{1}{2}; -1/\tau) \\
&= (\tau/i)^{1/2} \exp\{\pi i u^2 \tau\}\exp\{\pi i(u + \tfrac{1}{4})\tau\}\theta[{}^0_0](\tau u + \tau/2; \tau) \\
&= (\tau/i)^{1/2} \exp\{\pi i u^2 \tau\}\theta[{}^1_0](\tau u; \tau).
\end{aligned}$$

And also, replacing τ by $-1/\tau$ and u by τu in this last formula gives

$$\theta[{}^0_1](\tau u; \tau) = (i/\tau)^{1/2} \exp\{-\pi i u^2 \tau\}\theta[{}^1_0](u; -1/\tau).$$

Finally, using this last identity as well as the definitions of the theta functions, we compute

$$\begin{aligned}
\theta[{}^1_1](u; -1/\tau) &= \theta[{}^1_0](u + \tfrac{1}{2}; -1/\tau) \\
&= (\tau/i)^{1/2} \exp\{\pi i(u + \tfrac{1}{2})^2 \tau\}\theta[{}^0_1](\tau(u + \tfrac{1}{2}); \tau) \\
&= (\tau/i)^{1/2}(-i) \exp\{\pi i u^2 \tau\}\theta[{}^1_1](\tau u; \tau).
\end{aligned}$$

So, all together,

$$\theta[{}^\delta_\varepsilon](u; -1/\tau) = (\tau/i)^{1/2}(-i)^{\delta\varepsilon} \exp\{\pi i u^2 \tau\}\theta[{}^\varepsilon_\delta](u\tau; \tau). \qquad (3.54)$$

We also must compute $\theta[{}^\delta_\varepsilon](u; \tau + 1)$. These come almost directly from the definitions, specifically

$$\begin{aligned}
\theta[{}^1_\varepsilon](u; \tau + 1) &= \sum \exp\{\pi i[(n + \tfrac{1}{2})^2(\tau + 1) + 2(n + \tfrac{1}{2})(u + \varepsilon/2)]\} \\
&= e^{\pi i/4}\theta[{}^1_\varepsilon](u; \tau)
\end{aligned} \qquad (3.55)$$

and

$$\begin{aligned}
\theta[{}^0_\varepsilon](u; \tau + 1) &= \sum \exp\{\pi i[n^2(\tau + 1) + 2n(u + \varepsilon/2)]\} \\
&= \sum \exp\{\pi i[n^2(\tau) + 2n(u + (1 - \varepsilon)/2)]\} \qquad (3.56) \\
&= \theta[{}^{\ 0}_{1-\varepsilon}](u; \tau).
\end{aligned}$$

---

† The original reference for this formula, together with an interesting discussion of the origins of the material of this section, appears in Chapter 21, Section 5, of E. T. Whittaker and G. N. Watson's classic *Modern Analysis*, Cambridge: Cambridge University Press, 1927.

From these formulas we conclude, for example, that

$$\theta[^0_0](0;\tau)^8 + \theta[^1_0](0;\tau)^8 + \theta[^0_1](0;\tau)^8 \qquad (3.57)$$

is a modular form of weight 2 with value 2 at infinity; so by (3.51)

$$(3.57) = \frac{2\cdot 15}{\pi^4}\, c_2(\tau).$$

Similarly,

$$(\theta[^0_1]^4 + \theta[^0_0]^4)(\theta[^1_0]^4 + \theta[^0_0]^4)(\theta[^0_1]^4 - \theta[^1_0]^4)$$

must be

$$\frac{189}{\pi^6}\, c_4(\tau).$$

3.14 A Fundamental Domain for Γ_2

So certain polynomials in fourth powers of (even) theta-nulls generate the ring of modular forms.

Now let us restrict ourselves to the subgroup

$$\Gamma_2 \le \mathrm{Sp}_1(\mathbb{Z})$$

of those matrices which are the identity modulo 2. If we ignore, for the moment, conditions at the boundary, then a function $c(\tau)$ is a modular form of weight k with respect to Γ_2 if

$$c(\tau + 2) = c(\tau),$$

$$c\left(\frac{1}{2\tau + 1}\right) = (2\tau + 1)^{2k} c(\tau).$$

Recall that we showed before that $\Gamma_2/\{\pm \text{identity matrix}\}$ has generators

$$\begin{bmatrix} 1 & 0 \\ 2 & 1 \end{bmatrix} \quad \text{and} \quad \begin{bmatrix} 1 & 2 \\ 0 & 1 \end{bmatrix}.$$

Now

$$\begin{bmatrix} 1 & 0 \\ 2 & 1 \end{bmatrix} = \begin{bmatrix} 0 & 1 \\ -1 & 0 \end{bmatrix}\begin{bmatrix} 1 & -2 \\ 0 & 1 \end{bmatrix}\begin{bmatrix} 0 & -1 \\ 1 & 0 \end{bmatrix},$$

so that it is easy to check that

$$\theta[^\delta_\varepsilon](0,\tau)^4$$

is a modular form of weight 1 with respect to Γ_2 whenever $\delta\varepsilon = 0$. Now earlier in the chapter we showed that

$$\lambda(\tau) = -\theta[^1_0]^4/\theta[^0_1]^4$$

and that the mapping

$$\lambda: \Gamma_2 \backslash \mathcal{H} \to \mathbb{C}$$

is in fact injective. Indeed we have a diagram

$$
\begin{array}{ccc}
\Gamma_2 \mathcal{H} & \cong & \mathbb{C}\mathbb{P}_1 - \{0, 1, \infty\} \\
\Big\downarrow \tau \longmapsto & & \Big\downarrow (-\theta[^1_0]^4, \theta[^0_1]^4) \\
& & \text{algebraic map} \\
\mathcal{M} & \cong & \mathbb{C}\mathbb{P}_1 - \{\infty\}
\end{array}
\tag{3.58}
$$

Now the domain on $\mathcal{H}$ given by Figure 3.10 is made up of six fundamental domains for the action of $\mathrm{Sp}_1(\mathbb{Z})$ and so is easily seen to be a fundamental domain for the action of Γ_2. Now if $\delta \cdot \varepsilon = 0$, we have

$$
\begin{aligned}
\theta[^\delta_\varepsilon](0; \tau) &= \sum \exp\{\pi i[(n + \delta/2)^2\tau + 2(n + \delta/2)(\varepsilon/2)]\} \\
&= \sum \exp\{\pi i(n^2\tau + \delta n\tau + \delta\tau/4 + \varepsilon n)\}.
\end{aligned}
$$

So if $x > 0$, we obtain

$$\lim_{x \to \infty} \theta[^\delta_\varepsilon](0; ix) = (1 - \delta),$$

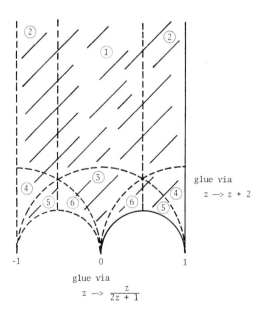

Figure 3.10. A fundamental domain for Γ_2 broken up into six fundamental domains for $\mathrm{Sp}_1(\mathbb{Z})$.

and so by (3.54)

$$\lim_{x\to 0^+} x^{1/2}\theta[^\delta_\varepsilon](0; ix) = \lim_{x\to\infty} \theta[^\varepsilon_\delta](0; ix) = (1-\varepsilon).$$

Finally using (3.55) and (3.56) gives

$$\lim_{x\to 0^+} x^{1/2}\theta[^1_0](0; 1+ix) = e^{\pi i/4},$$

$$\lim_{x\to 0^+} x^{1/2}\theta[^0_\varepsilon](0; 1+ix) = \varepsilon.$$

From these formulas we conclude that the coordinate function

$$-\theta[^1_0]^4/\theta[^0_1]^4$$

in (3.58) has the following limit values:

$$i\infty \longrightarrow 0,$$
$$i0^+ \longrightarrow \infty,$$
$$1 + i0^+ \longrightarrow 1.$$

Also the function

$$\theta[^0_0]^4/\theta[^0_1]^4$$

has the limit values

$$i\infty \longrightarrow 1,$$
$$i0^+ \longrightarrow \infty,$$
$$1 + i0^+ \longrightarrow 0.$$

Thus

$$1 - (-\theta[^1_0]^4/\theta[^0_1]^4) = (\theta[^0_0]^4/\theta[^0_1]^4);$$

in other words, we have *Riemann's theta relation:*

$$\theta[^0_1]^4 + \theta[^1_0]^4 = \theta[^0_0]^4. \tag{3.59}$$

3.15 Jacobi's Identity

Finally all the theta-nulls are related by a very nice identity, called *Jacobi's identity:*

$$\frac{\partial\theta[^1_1](u; \tau)}{\partial u}\bigg|_{u=0} = (-\pi)\theta[^0_0]\theta[^0_1]\theta[^1_0]. \tag{3.60}$$

To see this, raise both sides of equation (3.60) to the eighth power. The right-hand side is an automorphic form of weight 6 for $\text{Sp}_1(\mathbb{Z})$ with a zero at infinity (a so-called *cusp form*). This follows immediately from equations (3.54)–(3.56). Using equation (3.55) gives

$$\left.\frac{\partial\theta[^1_1](u;\,\tau+1)}{\partial u}\right|_{u=0} = e^{\pi i/4}\left.\frac{\partial\theta[^1_1](u;\,\tau)}{\partial u}\right|_{u=0},$$

and using (3.54) gives

$$\left.\frac{\partial\theta[^1_1](u;\,-1/\tau)}{\partial u}\right|_{u=0} = (-i)\left(\frac{\tau}{i}\right)^{1/2}\tau\,\left.\frac{\partial\theta[^1_1](u;\,\tau)}{\partial u}\right|_{u=0},$$

so that the left-hand side of (3.60), raised to the eighth power, is also a modular form of weight 6. But now computing directly gives

$$\left.\frac{\partial\theta[^1_1](u;\,\tau)}{\partial u}\right|_{u=0} = \sum \pi(2n+1)(-1)^{(n+1)}\exp\{\pi i\tau(n+\tfrac{1}{2})^2\},$$

so putting $\tau = ix$ and letting $x \mapsto +\infty$, we get that the left-hand side of (3.60) also has a zero at infinity. Since there is only one cusp form of weight 6 (up to multiplicative constant) by the formula on the orders of the zeros of a modular form given in (3.48)–(3.49), the equation (3.60) must be true up to a multiplicative constant. To evaluate the constant, compute the coefficient of $e^{\pi i\tau/4}$ on both sides of the equation.

CHAPTER FOUR

The Jacobian Variety

4.1 Cohomology of a Complex Curve

Again in this chapter we will assume a certain vague familiarity with the cohomology of sheaves and the theory of line bundles. This material is readily accessible in the modern literature, and some of it can be guessed once one is familiar with Čech cohomology of topological spaces. Our goal in this chapter is to examine in detail the *Jacobian variety* which is associated in an intrinsic way with each compact complex manifold of dimension 1. It is the existence of this auxiliary variety that makes the theory of compact manifolds of dimension 1 so much more beautiful and complete than the theory of complex manifolds of higher dimensions. The higher-dimensional theory often still consists of a struggle to resurrect or replace in some special case or other the one-dimensional theory.

To begin, let C be a compact complex manifold of dimension 1. C is then triangulable and thereby becomes a finite simplicial complex or a finite cell complex. It is not hard to show that as a cell complex C can be presented in the form shown in Figure 4.1, that is, a bouquet of $2g$ one-spheres $a_1, \beta_1, \alpha_2, \ldots, \beta_g$ with common point p to which there is attached a two-cell $\hat{C}$. Thus

$$H_0(C; \mathbb{Z}) = \mathbb{Z} \cdot \{p\},$$

$$H_1(C; \mathbb{Z}) = \sum_j (\mathbb{Z} \cdot \{\alpha_j\} + \mathbb{Z} \cdot \{\beta_j\}),$$

$$H_2(C; \mathbb{Z}) \cong \mathbb{Z} \cdot \{\hat{C}\}.$$

The integer g is called the *genus* of C. Also it is clear from Figure 4.1 that

$$\alpha_j \cdot \beta_k = \text{Kronecker } \delta_{jk},$$

$$\alpha_j \cdot \alpha_k = 0,$$

$$\beta_j \cdot \beta_k = 0.$$

Because of these formulas we call $\{\alpha_j, \beta_k\}$ a *symplectic* basis of $H_1(C; \mathbb{Z})$.

Let us now begin with the exact sheaf sequence on C:

$$0 \longrightarrow \mathbb{Z} \longrightarrow \mathcal{O} \xrightarrow{e^{2\pi i(\ )}} \mathcal{O}^* \longrightarrow 0 \qquad (4.1)$$

where $\mathcal{O}$ is the sheaf of holomorphic functions on C (which we have already seen in Chapter Two), and $\mathcal{O}^*$ is the sheaf of *nowhere-zero* holomorphic functions on C whose first cohomology group

$$H^1(C; \mathcal{O}^*)$$

is exactly the group of holomorphic equivalence classes of holomorphic line bundles on C. We will build the Jacobian variety of C from a piece of the cohomology sequence associated with the sheaf sequence (4.1), namely, the part

$$0 \to H^1(C; \mathbb{Z}) \to H^1(C; \mathcal{O}) \to H^1(C; \mathcal{O}^*) \to H^2(C; \mathbb{Z}) \to 0. \qquad (4.2)$$

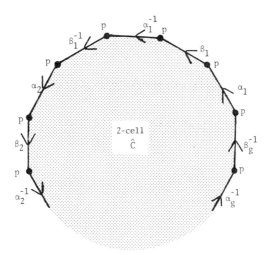

Figure 4.1. Cutting open a compact complex one-dimensional manifold in order to lay it out in the plane.

Then we define the *Picard variety* of C:

$$\text{Pic}^0(C) = \frac{H^1(C; \mathcal{O})}{H^1(C; \mathbb{Z})}$$

$$= \text{kernel}(H^1(C; \mathcal{O}^*) \to H^2(C; \mathbb{Z})). \qquad (4.3)$$

Even to write the sequence (4.3) we must assume many things about the cohomology of the sheaf $\mathcal{O}$ on C, in particular that

$$H^n(C; \mathcal{O}) = 0$$

whenever $n > \dim_{\mathbb{C}} C = 1$. This fact, as well as many others which we will use without proof, can be found (with proof) in Gunning [3]. These facts center around the study of the double complex of sheaves

$$
\begin{array}{ccccccc}
\uparrow & & \uparrow & & \uparrow & & \\
\mathcal{A}^{0,2} & \xrightarrow{\partial} & \mathcal{A}^{1,2} & \xrightarrow{\partial} & \mathcal{A}^{2,2} & \xrightarrow{\partial} & \\
\uparrow \bar{\partial} & & \uparrow \bar{\partial} & & \uparrow & & \\
\mathcal{A}^{0,1} & \xrightarrow{\partial} & \mathcal{A}^{1,1} & \xrightarrow{\partial} & \mathcal{A}^{2,1} & \xrightarrow{\partial} & \\
\uparrow \bar{\partial} & & \uparrow \bar{\partial} & & \uparrow & & \\
\mathcal{A}^{0,0} & \xrightarrow{\partial} & \mathcal{A}^{1,0} & \xrightarrow{\partial} & \mathcal{A}^{2,0} & \xrightarrow{\partial} & \\
\uparrow & & \uparrow & & \uparrow & & \\
0 \longrightarrow \mathbb{C} \longrightarrow \mathcal{O} & \xrightarrow{\partial} & \Omega^1 & \xrightarrow{\partial} & \Omega^2 & \xrightarrow{\partial} &
\end{array}
\qquad (4.4)
$$

where

$\mathcal{A}^{i,j}$ = sheaf of C^∞ complex-valued differential forms of type (i, j) on C,

Ω^j = sheaf of holomorphic j-forms on C.

($\mathcal{O} = \Omega^0$.) The point is that the bigraded complex

$$\left(\sum \mathcal{A}^{i,j}, \partial + \bar{\partial} \right)$$

is a finite resolution of the constant sheaf $\mathbb{C}$, and therefore its global section can be used to compute the usual deRham cohomology groups

$$H^*(C; \mathbb{C}).$$

Much deeper is the consequence of Hodge theory for Kähler manifolds that the spectral sequence associated with the filtration

$$F(p) = \sum_{i \geq p} \mathcal{A}^{i,j}$$

of the complex (4.4) "degenerates at E_1." In other words,

$$H^0(C; \mathbb{C}) = H^0(C; \mathcal{O}),$$
$$H^1(C; \mathbb{C}) \cong H^1(C; \mathcal{O}) \oplus H^0(C; \Omega^1).$$

4.2 Duality

The pairing

$$H^1(M; \mathbb{C}) \times H^1(M; \mathbb{C}) \longrightarrow \mathbb{C},$$

$$(\omega, \eta) \longmapsto \int_C \omega \wedge \eta \qquad (4.5)$$

is nondegenerate and skew-symmetric because it is the complexification of the cup-product pairing on integral cohomology. If

$$\{d\xi_j, d\eta_k\}_{j, k = 1, \dots, g} \qquad (4.6)$$

is the basis of $H^1(C; \mathbb{Z})$ such that

$$\alpha_j \cdot \gamma = \int_\gamma d\xi_j,$$

$$\beta_k \cdot \gamma = \int_\gamma d\eta_k,$$

we call this basis the *Poincaré dual basis* of the basis $\{\alpha_j, \beta_k\}$ of $H_1(C; \mathbb{Z})$. Now if

$$\{\omega_1, \dots, \omega_g\}$$

is a basis for $H^0(C; \Omega^1)$, then we have that

$$\bar{\omega}_1, \dots, \bar{\omega}_g$$

must be also closed, linearly independent (0, 1) forms. More precisely, they represent linearly independent elements of $H^1(C; \mathbb{C})$, and since they lie in

$$\frac{\bar{\partial}\text{-closed } (0, 1)\text{-forms}}{\bar{\partial}\text{-exact } (0, 1)\text{-forms}} = H^1(C; \mathcal{O}) \subseteq H^1(C; \mathbb{C}),$$

they must, for dimensional reasons, be a basis of this space. Thus the deRham group splits naturally:

$$H^1(C; \mathbb{C}) = (\text{subspace spanned by } \{\omega_1, \dots, \omega_g\})$$

$$+ (\text{subspace spanned by } \{\bar{\omega}_1, \dots, \bar{\omega}_g\})$$

$$\underset{\text{def}}{=} H^{1, 0}(C) + H^{0, 1}(C).$$

From this it is easy to see that the matrix

$$\Sigma = \int_{\beta_k} \omega_j$$

is of maximal rank so that for appropriate choice of $\omega_1, \ldots, \omega_g$, we can effect that

$$\Sigma = (\text{identity matrix})$$

From now on we will assume that the ω_j are so chosen; then we have

$$\begin{bmatrix} \omega_1 \\ \vdots \\ \omega_g \end{bmatrix} = [I \quad \Omega] \begin{bmatrix} d\xi_1 \\ \vdots \\ d\eta_g \end{bmatrix}, \tag{4.7}$$

where Ω is a complex-valued $g \times g$ matrix of maximal rank.

The formulas

$$\int_C \omega_j \wedge \omega_k = 0$$

imply that

$$\Omega = {}^t\Omega \tag{4.8}$$

(called the *first Riemann relation*), and the formulas

$$\left(i \int \omega_j \wedge \bar{\omega}_j \right) > 0$$

imply that

$$G = (\text{imaginary part } \Omega) = \text{positive definite symmetric matrix} \tag{4.9}$$

(called the *second Riemann relation*; we have already computed a special case of it in Chapter Two).

Consequently we have the formula

$$\begin{bmatrix} I & \Omega \\ I & \bar{\Omega} \end{bmatrix} \begin{bmatrix} 0 & I \\ -I & 0 \end{bmatrix} \begin{bmatrix} I & I \\ {}^t\Omega & {}^t\bar{\Omega} \end{bmatrix} \begin{bmatrix} 0 & I \\ -I & 0 \end{bmatrix} = 2i \begin{bmatrix} G & 0 \\ 0 & G \end{bmatrix}.$$

This shows that the column vectors of the matrix

$$[I \quad \Omega]$$

are linearly independent over $\mathbb{R}$, since otherwise the matrix

$$\begin{vmatrix} I & \Omega \\ I & \bar{\Omega} \end{vmatrix}$$

could not possibly be of rank $2g$.

4.3 The Chern Class of a Holomorphic Line Bundle

We are now in a position to begin to understand the structure of the Picard variety which we defined at the beginning of this chapter. From the complex (4.4) we see that the mapping

$$H^1(C; \mathbb{Z}) \to H^1(C; \mathcal{O})$$

is simply the restriction of the natural projection

$$
\begin{array}{c}
H^1(C; \mathbb{C}) \longrightarrow H^1(C; \mathcal{O}), \\
\sum a_j \omega_j + b_j \bar{\omega}_j \longmapsto \sum b_j \bar{\omega}_j.
\end{array}
\tag{4.10}
$$

Under this projection the mapping

$$H^1(C; \mathbb{R}) \to H^1(C; \mathcal{O})$$

is an isomorphism of real vector spaces, since a differential form is real if and only if it is its own conjugate, that is, if and only if it has the form

$$\sum a_j \omega_j + \bar{a}_j \bar{\omega}_j.$$

Thus we have a natural isomorphism

$$\mathrm{Pic}^0(C) \cong H^1(C; \mathbb{R})/H^1(C; \mathbb{Z}),
\tag{4.11}$$

where the complex structure is induced on this real $2g$-dimensional torus by means of the mapping (4.10).

The next step in understanding (4.3) is to remind ourselves that if we consider the group

$$H^1(C; \mathcal{O}^*)$$

as computed via Čech cohomology of sheaves, then it is simply the group of holomorphic equivalence classes of complex holomorphic line bundles on C. Topologically these line bundles are classified by their *Chern class*, an element of $H^2(C; \mathbb{Z}) \cong \mathbb{Z}$. (See, for example, Steenrod [10].) We ought to check that the map

$$H^1(C; \mathcal{O}^*) \to H^2(C; \mathbb{Z})
\tag{4.12}$$

in the sequence (4.2) is indeed the Chern class mapping. Then we will be able to conclude that $\mathrm{Pic}^0(C)$ is simply the moduli space of complex holomorphic line bundles on C which are topologically trivial.

So let's analyze the definition of the mapping (4.12). We will use the fact from Gunning's Riemann surfaces book [3] that every line bundle on C is a product of bundles of the type

L_p = holomorphic bundle associated to the divisor consisting of one point $p \in C$

or L_p^{-1}. So we will prove that, with appropriate choice of sign in (4.12),

$$\{L_p\} \mapsto +1.$$

Let us use a covering

$$U, V_1, V_2$$

of C which near p looks like that shown in Figure 4.2. If z is an analytic coordinate on U with $z(p) = 0$, then

$$z \quad \text{on } U,$$

$$1 \quad \text{on } V_1 \text{ and } V_2$$

give a global section of L_p, that is, the pasting function ϕ_{UV_j} used to put

$$U \times \mathbb{C} \quad \text{and} \quad V_j \times \mathbb{C}$$

together to make L_p is

$$\phi_{UV_j}\colon \quad U \cap V_j \longrightarrow \mathbb{C}^*$$
$$q \longmapsto 1/z(q)$$

and $\phi_{V_1V_2} = 1$. The first step in computing the image of

$$\{\phi_{UV_1}, \phi_{UV_2}, \phi_{V_1V_2}\} \in H^1(C; \mathcal{O}^*) \tag{4.13}$$

in $H^2(C; \mathbb{Z})$ is to select a Čech one-cochain with coefficients in $\mathcal{O}$ which maps to (4.13) under the mapping

$$\mathcal{O} \longrightarrow \mathcal{O}^*$$
$$\sigma \longmapsto e^{2\pi i \sigma}.$$

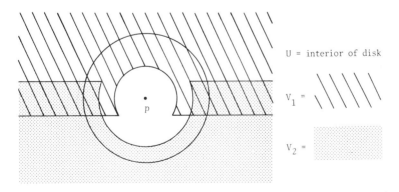

Figure 4.2. The three sets of a Čech covering which computes the Chern class of L_p.

We use

$$\left|\frac{\log \phi_{UV_1}}{2\pi i}, \frac{\log \phi_{UV_2}}{2\pi i}, 0\right| \tag{4.14}$$

where the branches of $\log \phi_{UV_1}$ and $\log \phi_{UV_2}$ are chosen to coincide along z real, negative.

We next compute the Čech coboundary of (4.14):

$$\delta(4.14)_{UV_1V_2} = \frac{1}{2\pi i}\left(\log \phi_{UV_1} - \log \phi_{UV_2}\right)$$

$$= \left|\begin{matrix}1\\0\end{matrix} \text{ on } \left|\begin{matrix}\text{right-hand}\\\text{left-hand}\end{matrix}\right| \text{component of } U \cap V_1 \cap V_2. \right.$$

Clearly, then, this element represents the generator of the Čech cohomology group

$$H^2(C; \mathbb{Z}).$$

To compare Čech and deRham cohomology, we use the diagram of complexes:

$$\begin{array}{ccccccccc}
& \uparrow\delta & & \uparrow\delta & & \uparrow\delta & & & \\
0 \longrightarrow & C^2(C;\mathbb{C}) & \longrightarrow & C^2(C;\mathscr{A}^0) & \xrightarrow{d} & C^2(C;\mathscr{A}^1) & \xrightarrow{d} & & \\
& \uparrow\delta & & \uparrow\delta & & \uparrow\delta & & & \\
0 \longrightarrow & C^1(C;\mathbb{C}) & \longrightarrow & C^1(C;\mathscr{A}^0) & \xrightarrow{d} & C^1(C;\mathscr{A}^1) & \xrightarrow{d} & & (4.15) \\
& \uparrow\delta & & \uparrow\delta & & \uparrow\delta & & \uparrow & \\
0 \longrightarrow & C^0(C;\mathbb{C}) & \longrightarrow & C^0(C;\mathscr{A}^0) & \xrightarrow{d} & C^0(C;\mathscr{A}^1) & \xrightarrow{d} & C^0(C;\mathscr{A}^2) & \xrightarrow{d} \\
& \uparrow & & \uparrow & & \uparrow & & \uparrow & \\
0 \longrightarrow & \mathbb{C} & \longrightarrow & \Gamma\mathscr{A}^0 & \xrightarrow{d} & \Gamma\mathscr{A}^1 & \xrightarrow{d} & \Gamma\mathscr{A}^2 & \xrightarrow{d} \\
& \uparrow & & \uparrow & & \uparrow & & \uparrow & \\
& 0 & & 0 & & 0 & & 0 &
\end{array}$$

We can put a metric on the line bundle

$$\{\xi_{\alpha\beta}\} \in H^1(C; \mathcal{O}^*)$$

by finding functions

$$r_\alpha: U_\alpha \to \mathbb{R}_+$$

such that

$$r_\alpha = r_\beta |\xi_{\alpha\beta}|^2. \tag{4.16}$$

Then

$$\{\xi_{\alpha\beta}\} \in H^1(C; \mathcal{O}) \mapsto \{\log \xi_{\alpha\beta}\} \in C^1(C; \mathcal{O}) \mapsto$$

$$\delta\left\{\frac{1}{2\pi i} \log \xi_{\alpha\beta}\right\} \in C^2(C; \mathbb{Z}) \subseteq C^2(C; \mathcal{A}^0).$$

This last element represents the same class in $H^2(C; \mathbb{C})$ as does

$$\left\{\frac{-1}{2\pi i} \partial \log \xi_{\alpha\beta}\right\} \in C^1(C; \mathcal{A}^1),$$

since their difference is a coboundary in the double complex in (4.15). But by (4.16)

$$\partial \log \xi_{\alpha\beta} = \partial \log r_\alpha - \partial \log r_\beta$$

so that the same class is represented by

$$\left\{\frac{1}{2\pi i} \partial\bar\partial \log r_\alpha\right\} \in \Gamma\mathcal{A}^2 \subseteq C^0(C; \mathcal{A}^2).$$

So, for example, in the case of L_p we can use

$$r_U = \begin{cases} 1/|z|^2 & \text{near the edge of } U, \\ \text{"smoothed-off" positive function near the center,} \end{cases}$$

$$r_{V_k} = 1.$$

Thus

$$\int_U \partial\bar\partial \log r_U = \int_{\partial U} \bar\partial \log r_U = \int_{\partial U} \overline{(-d \log z)} = 2\pi i,$$

and we see indeed that

$$\frac{1}{2\pi i} \{\partial\bar\partial \log r_\alpha\} \in \Gamma\mathcal{A}^2$$

is a deRham representative of the Chern class of L_p.

From the preceding considerations it is also clear that if a line bundle has constant transition functions $\phi_{\alpha\beta}$, then its Chern class is zero. Conversely, the commutative sheaf diagram

$$\begin{array}{ccccccccc}
0 & \longrightarrow & \mathbb{Z} & \longrightarrow & \mathbb{C} & \longrightarrow & \mathbb{C}^* & \longrightarrow & 0 \\
& & \| & & \cap & & \cap & & \\
0 & \longrightarrow & \mathbb{Z} & \longrightarrow & \mathcal{O} & \longrightarrow & \mathcal{O}^* & \longrightarrow & 0
\end{array}$$

gives a commutative cohomology diagram

$$
\begin{array}{ccccc}
H^1(C;\mathbb{C}) & \longrightarrow & H^1(C;\mathbb{C}^*) & \longrightarrow & H^2(C;\mathbb{Z}) \\
\downarrow \alpha & & \downarrow & & \| \\
H^1(C;\mathcal{O}) & \longrightarrow & H^1(C;\mathcal{O}^*) & \longrightarrow & H^2(C;\mathbb{Z})
\end{array}
$$

and since the mapping α has been shown to be surjective, it is clear that a line bundle that has Chern class zero can be constructed using constant transition functions.

4.4 Abel's Theorem for Curves

Next we recall the *Lefschetz duality theorem* in algebraic topology that there is a nondegenerate pairing

$$
H_1(M;\mathbb{C}) \otimes H^1(M;\mathbb{C}) \longrightarrow \mathbb{C}
$$

$$
(\gamma, \phi) \longmapsto \int_\gamma \phi \tag{4.17}
$$

for any compact manifold M. This pairing, restricted to *integral* homology and cohomology is integral and unimodular. Define

$$
\begin{aligned}
H_{1,0}(C) &= (H^{0,1}(C))^\perp, \\
H_{0,1}(C) &= (H^{1,0}(C))^\perp,
\end{aligned}
\tag{4.18}
$$

where the superscript $\perp$ means the annihilator of the subspace in question with respect to the pairing (4.17). Then we have a decomposition

$$
H_1(C;\mathbb{C}) = H_{1,0}(C) + H_{0,1}(C)
$$

into complex subspaces, and as before the projection

$$
H_1(C;\mathbb{R}) \to H_{1,0}(C) \tag{4.19}
$$

is an isomorphism. Also we can identify

$$
H_{1,0}(C) = H^{1,0}(C)^*,
$$

the dual complex vector space of $H^{1,0}(C)$, via the pairing (4.17). We define

$$
\mathrm{Alb}(C) = \frac{H_1(C;\mathbb{R})}{H_1(C;\mathbb{Z})} \tag{4.20}
$$

and give it a complex structure via the projection mapping (4.19).

Now it will turn out that, as complex tori,

$$
\mathrm{Alb}(C) \cong \mathrm{Pic}^0(C). \tag{4.21}
$$

(This "common" variety is called the *Jacobian variety*.) The isomorphism (4.21) is a disguised form of the classical *Abel's theorem* (which we saw in Section 3.3 in the case $g = 1$). Before seeing why all this is true we need some further constructions.

First, if we pick a basepoint $p_0 \in C$, we can map

$$\kappa: \quad C \longrightarrow \text{Alb}(C)$$

$$p \longmapsto \left(\int_{p_0}^{p} \right). \tag{4.22}$$

This map makes sense because if we pick a path γ joining p_0 to p, then

$$\int_{\gamma} : H^1(C; \mathbb{R}) \to \mathbb{R}$$

is well defined so that $\int_{\gamma}$ is indeed an element of $H_1(C; \mathbb{R})$ via the pairing (4.17). Now if we use a different path from p_0 to p, we get perhaps a new element of $H_1(C; \mathbb{R})$, but it differs from the old one by an element of $H_1(C; \mathbb{Z})$, so the mapping κ is well defined! To check that it is complex analytic, notice that if $\omega \in H^{1, 0}(C)$, then

$$\int_{p_0}^{p} \omega$$

is an analytic function of p. But also the projection of $\int_{p_0}^{p}$ into $H_{1, 0}(C)$ is precisely the functional

$$\omega \mapsto \int_{p_0}^{p} \omega$$

in $H^{1, 0}(C)^*$. Also it is immediate to check (as we did at the beginning of Chapter Three for cubic curves) that we have isomorphisms

$$\kappa_*: \quad H_1(C; \mathbb{Z}) \longrightarrow H_1(\text{Alb}(C); \mathbb{Z}),$$
$$\kappa^*: H^1(\text{Alb}(C); \mathbb{Z}) \longrightarrow H^1(C; \mathbb{Z}). \tag{4.23}$$

We have chosen bases for integral homology and cohomology on C. We shall use the same symbols to denote the corresponding bases on $\text{Alb}(C)$.

Now we want to construct an isomorphism between $\text{Alb}(C)$ and $\text{Pic}^0(C)$. If we ignore their complex structures, this is an easy task. For example, the intersection pairing on $H_1(C; \mathbb{Z})$ gives the Poincaré mapping

$$H_1(C; \mathbb{Z}) \longrightarrow H^1(C; \mathbb{Z}),$$

$$\alpha \longmapsto (\alpha' \mapsto \alpha \cdot \alpha'). \tag{4.24}$$

[Recall (4.17) which makes H^1 the dual of H_1.] Since (4.24) is an isomorphism, it clearly induces an isomorphism

$$\text{Alb}(C) = \frac{H_1(C; \mathbb{R})}{H_1(C; \mathbb{C})} \cong \frac{H^1(C; \mathbb{R})}{H^1(C, \mathbb{Z})} = \text{Pic}^0(C). \tag{4.25}$$

The crucial point to check, of course, is whether this mapping (4.25) is complex analytic.

To check analyticity we rewrite the map (4.25). Let

$$\Phi = \sum d\xi_j \wedge d\eta_j, \tag{4.26}$$

which can be considered a deRham two-form on $\text{Alb}(C)$ or a skew-symmetric bilinear form on $H_1(C; \mathbb{R})$. In its former role we can use it to define the topological contraction mapping

$$H_1(C; \mathbb{R}) = H_1(\text{Alb}(C); \mathbb{R}) \longrightarrow H^1(\text{Alb}(C); \mathbb{R}) = H^1(C; \mathbb{R}),$$

$$\alpha \longmapsto \langle \alpha, \Phi \rangle.$$

The corresponding mapping at the level of simplicial complexes is the "cap product with Φ" mapping, but here we *ring* isomorphisms

$$H_*(\text{Alb}(C); \ ) \cong \Lambda^* H_1(\text{Alb}(C); \ ),$$
$$H^*(\text{Alb}(C); \ ) \cong \Lambda^* H^1(\text{Alb}(C); \ ), \tag{4.27}$$

where the product in $H_*(\text{Alb}(C); \ )$ is the *Pontryagin product*, which can be defined in an abelian Lie group G as follows:

If M_j is a compact r_j-manifold and

$$\alpha_j \colon M_j \to G$$

is a continuous map, then the product of the elements

$$\alpha_j(M_j) \in H_{r_j}(G)$$

is the image of the distinguished generator of $H_{\Sigma r_j}(\Pi_j M_j)$ in $H_{\Sigma r_j}(G)$ under the mapping

$$\Pi_j M_j \longrightarrow G,$$

$$(x_j) \longmapsto \Pi_j \alpha_j(x_j).$$

So by (4.27) the map $\alpha \mapsto \langle \alpha, \Phi \rangle$ can be understood as the contraction mapping which is always defined between the wedge algebra of a vector space and the wedge algebra of its dual. Then we check

$$\langle \alpha_j, \sum d\xi_j \wedge d\eta_j \rangle = \langle \alpha_j, -\sum d\eta_j \wedge d\xi_j \rangle$$
$$= -\langle \alpha_j, d\eta_j \rangle \, d\xi_j$$
$$= d\xi_j,$$

so that the contraction mapping is the same as the mapping (4.24).

Now let us rewrite the differential form Φ in terms of the basis

$$\{\omega_1, \ldots, \omega_g, \bar{\omega}_1, \ldots, \bar{\omega}_g\}$$

of $H^1(M; \mathbb{C})$. By (4.7) and the matrix formula following the definition of G in (4.9)

$$\begin{bmatrix} d\xi_1 \\ \vdots \\ d\eta_g \end{bmatrix} = -\frac{i}{2} \begin{bmatrix} -\bar{\Omega} & \Omega \\ I & -I \end{bmatrix} \begin{bmatrix} G^{-1} & 0 \\ 0 & G^{-1} \end{bmatrix} \begin{bmatrix} \omega_1 \\ \vdots \\ \bar{\omega}_g \end{bmatrix}.$$

This equation can be thought of either in $H^1(C; \mathbb{C})$ or in $H^1(\mathrm{Alb}(C); \mathbb{C})$. So we obtain in $H^2(\mathrm{Alb}(C); \mathbb{C})$ the following expression for Φ:

$$[d\xi_1, \ldots, d\xi_g] \wedge \begin{bmatrix} d\eta_1 \\ \vdots \\ d\eta_g \end{bmatrix}$$

$$= -\frac{1}{4}[\omega_1 \quad \cdots \quad \bar{\omega}_g] \begin{bmatrix} -G^{-1} \cdot \bar{\Omega} \\ G^{-1} \cdot \Omega \end{bmatrix} \wedge [G^{-1} \quad -G^{-1}] \begin{bmatrix} \omega_1 \\ \vdots \\ \bar{\omega}_g \end{bmatrix}$$

$$= -\frac{1}{4}[\omega_1 \quad \cdots \quad \omega_g] \wedge [G^{-1}\bar{\Omega}G^{-1} - G^{-1}\Omega G^{-1}] \begin{bmatrix} \bar{\omega}_1 \\ \vdots \\ \bar{\omega}_g \end{bmatrix}$$

$$= \frac{i}{2}[\omega_1 \quad \cdots \quad \omega_g] \wedge G^{-1} \begin{bmatrix} \bar{\omega}_1 \\ \vdots \\ \bar{\omega}_g \end{bmatrix}.$$

We say that Φ is a two-form of type $(1, 1)$ on $\mathrm{Alb}(C)$. The fact that G^{-1} is positive definite and Φ is an integral cycle of type $(1, 1)$ makes $\mathrm{Alb}(C)$ into a Kähler manifold. (See for example, Hirzebruch [4], p. 123.)

Just as we have decomposed

$$H_1(C; \mathbb{C}) \qquad \text{and} \qquad H^1(C; \mathbb{C})$$

in subspaces $H_{1, 0}$, etc., according to type, we can also decompose the vector space

$$\mathrm{Hom}_{\mathbb{C}}(H_1(C; \mathbb{C}), H^1(C; \mathbb{C}))$$
$$\|$$
$$H^1(C; \mathbb{C}) \otimes H^1(C; \mathbb{C})$$
$$\|$$
$$H^{1, 0} \otimes H^{1, 0} + H^{1, 0} \otimes H^{0, 1} + H^{0, 1} \otimes H^{1, 0} + H^{0, 1} \otimes H^{0, 1}.$$

Also we can decompose the cohomology of $\mathrm{Alb}(C)$ into types using (4.27). This turns out to be the same thing as the *Hodge decomposition* on the Kähler manifold $\mathrm{Alb}(C)$. Suppose we are given an element

$$\phi \in H^{1,1}(\mathrm{Alb}(C)) \cap H^2(\mathrm{Alb}(C); \mathbb{Z}).$$

Then $\bar{\phi} = \phi$ and so

$$\phi = \sum c_{jk}\, \omega_j \wedge \bar{\omega}_k$$

with $[c_{jk}]$ a skew-Hermitian matrix. But the natural isomorphism on $\mathrm{Alb}(C)$,

$$H^{1,0} \otimes H^{0,1} \xrightarrow{\;\cong\;} H^{1,1}$$

$$\omega_j \otimes \bar{\omega}_k \longmapsto \omega_j \wedge \bar{\omega}_k,$$

says that ϕ determines a unique element

$$\phi^{1,0} = \sum c_{jk}\, \omega_j \otimes \bar{\omega}_k \in H^{1,0} \otimes H^{0,1}.$$

Next we put $\phi^{0,1} = \overline{\phi^{1,0}}$ and claim that

$$(\phi^{1,0} + \phi^{0,1})(\alpha \otimes \beta) = \langle \phi, \alpha \wedge \beta \rangle \tag{4.28}$$

for all $\alpha, \beta \in H_1(\mathrm{Alb}(C); \mathbb{C})$. To check that $\phi^{1,0} + \overline{\phi^{1,0}}$ has the desired property (4.28), we compute

$$(\phi^{1,0} + \phi^{0,1})(\alpha \otimes \beta) = \langle \sum c_{jk}\, \omega_j \otimes \bar{\omega}_k + \sum \bar{c}_{jk}\, \bar{\omega}_j \otimes \omega_k, \alpha \otimes \beta \rangle$$

$$= \langle \sum c_{jk}\, \omega_j \otimes \bar{\omega}_k, \alpha \otimes \beta - \beta \otimes \alpha \rangle$$

since $[c_{jk}]$ is skew-Hermitian. Thus, since $\phi \in H^2(\mathrm{Alb}(C); \mathbb{Z})$, $(\phi^{1,0} + \phi^{0,1})$ is actually an element of

$$H^1(\mathrm{Alb}(C); \mathbb{Z}) \otimes H^1(\mathrm{Alb}(C); \mathbb{Z}),$$

that is, corresponds to a homomorphism

$$H_1(C; \mathbb{Z}) \to H^1(C; \mathbb{Z}). \tag{4.29}$$

Again using (4.28), it is immediate that if $\phi = \Phi$ as in (4.26), then the last map sends

$$\alpha_j \mapsto d\xi_j, \qquad \beta_j \mapsto d\eta_j$$

and so is the mapping (4.24). It is also easy to check that (4.29) is an isomorphism if and only if the form ϕ, considered a skew-symmetric bilinear form on $H_1(\mathrm{Alb}(C); \mathbb{Z})$, is unimodular.

Finally we can check the analyticity of any mapping obtained by tensoring with $\mathbb{R}$ a map (4.29) coming from a form ϕ of type $(1, 1)$. Indeed the composition

$$H_{1,0} \to H_1(C; \mathbb{R}) \to H^1(C; \mathbb{R}) \to H^{0,1}$$

is given by the element $\phi^{1,0} \in \text{Hom}_{\mathbb{C}}(H_{1,0}, H^{0,1})$ and so is complex linear and therefore analytic. We therefore conclude

$$\text{Alb}(C) \cong \text{Pic}^0(C), \tag{4.30}$$

the isomorphism being given by the $(1, 1)$-form Φ. The common object in (4.30) is called the *Jacobian variety* of C and is denoted by

$$J(C).$$

4.5 The Classical Version of Abel's Theorem

To see how this relates to the classical Abel's theorem, notice that we have a diagram

$$
\begin{array}{ccc}
 & \xrightarrow{\ \kappa\ } & \text{Alb}(C) \\
C & & \downarrow \Phi \\
 & \xrightarrow{\ \mu\ } & \text{Pic}^0(C)
\end{array}
\tag{4.31}
$$

where $\mu(p)$ is the equivalence class of the line bundle given by the divisor $p - p_0$ and κ is as in (4.22). The usual form of Abel's theorem is then simply the assertion that the diagram (4.31) is commutative. We now proceed to prove this.

Let z be a holomorphic coordinate function on a disk U_0 in C; let

$$q_1, q_2 \in U_0,$$

and let

$$\{U_0\} \cup \{U_\alpha\}_{\alpha=1,\ldots,m}$$

be a finite cover of C such that if $\alpha \neq 0$ then U_α avoids a neighborhood of a simple path connecting q_1 and q_2 in U_0 (Figure 4.3). Now using Section 4.3 we see that the line bundle $L_{q_1} \otimes L_{q_2}^{-1}$ is represented in $H^1(C; \mathcal{O})$ by the cocycle $\{\phi_{\alpha\beta}\}$ where

$$\phi_{0\alpha} = \frac{1}{2\pi i}\{\log[z - z(q_2)] - \log[z - z(q_1)]\},$$

$$\phi_{\alpha\beta} = 0 \quad \text{if} \quad \alpha \neq 0, \beta \neq 0. \tag{4.32}$$

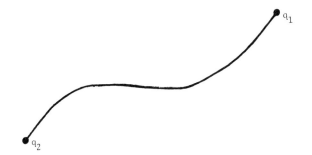

Figure 4.3. This is not a snake; it is a path from the point q_2 to the point q_1.

In fact, by the Riemann–Roch theorem [see (4.48)], there exists a meromorphic differential

$$\psi \tag{4.33}$$

on C such that the only poles of ψ are at q_1 and q_2 and those poles are simple with residues 1 and -1 respectively. So we might as well take z so that

$$\exp\left|2\pi i \int_{p_0}^{p} \psi\right| = \frac{z(p) - z(q_1)}{z(p) - z(q_2)}.$$

Next we can, and will, assume that elements of our chosen basis

$$d\xi_j, \, d\eta_k$$

of $H^1(C; \mathbb{Z})$ are such that they are identically zero in a neighborhood of $\bar{U}_0$ on C, and that the homology cycles α_j, β_k also avoid $\bar{U}_0$. Define

$$\psi^1 = \sum \left(\int_{\beta_j} \psi\right) d\xi_j - \sum \left(\int_{\alpha_j} \psi\right) d\eta_j.$$

Then off the path in Figure 4.3,

$$(\psi - \psi^1) = d\sigma$$

for some C^∞-function σ. Next let

$$\varepsilon_j(p) = \int_{p_0}^{p} \omega_j$$

be defined in a neighborhood of $\bar{U}_0$.

Now we have vector equations

$$\left(\int_{q_2}^{q_1} \omega_j \right) = (\varepsilon_j(q_1) - \varepsilon_j(q_2)) = \left(\int_{\partial U_0} \varepsilon_j \psi \right) = \left(\int_{\partial U_0} \varepsilon_j (\psi - \psi^1) \right)$$

$$= -\left(\int_{\partial U_0} \omega_j \sigma \right) = \left(\int_C \omega_j \wedge \psi^1 \right) = \left(\int_C \varepsilon \wedge \omega_j \right), \qquad (4.34)$$

where $\varepsilon = -(0, 1)$-component of ψ^1. On the other hand, ε represents the class (4.32) in $H^1(C; \mathcal{O})$, since, if we define

$$\sigma_\alpha = \sigma|_{U_\alpha}, \qquad \alpha \neq 0,$$

$$\sigma_0 = 0,$$

then

$$\bar{\partial}\{\sigma_\alpha\} = \varepsilon$$

and

$$\delta\{\sigma_\alpha\} = -\{\phi_{\alpha\beta}\}$$

[see (4.32)]. But now the Poincaré dual mapping (4.24) sends $\int_{q_2}^{q_1}$ to that differential form ρ such that

$$\int_{q_2}^{q_1} \eta = \int_C \rho \wedge \eta.$$

Thus by the formula (4.34), the element $\int_{q_2}^{q_1}$ must go, via (4.24), to

$$(\bar{\varepsilon} + \varepsilon).$$

Thus the mapping

$$\Phi \colon \mathrm{Alb}(C) \to \mathrm{Pic}^0(C)$$

takes $\int_{q_2}^{q_1}$ to the class of the bundle $L_{q_1} \otimes L_{q_2}^{-1}$, and the proof of Abel's theorem is complete.

By this theorem, if

$$\sum \int_{q_r}^{p_r} = 0 \qquad \text{in } \mathrm{Alb}(C),$$

then we must have a rational function f on C whose divisor

$$(f) = \sum p_r - \sum q_r.$$

To produce such a function explicitly let

$$\psi = \sum \psi_r$$

have poles only at $\{p_r\} \cup \{q_r\}$, residues $+1$ at p_r, -1 at q_r, etc., as in (4.33).

Then

$$f(p) = \exp\left|2\pi i \int_{p_0}^{p} \psi\right|$$ (4.35)

may not be a well-defined function because

$$\int_{\alpha_j} \psi, \qquad \int_{\beta_j} \psi$$ (4.36)

may not be integers. However we can adjust ψ by a sum

$$\sum a_j \omega_j$$

such that for all j,

$$\int_{\beta_j} \psi = 0.$$

So, as before, we put

$$\psi^1 = -\sum \left(\int_{\alpha_k} \psi\right) d\eta_k$$

and conclude that there are *integers* $m_1, \ldots, m_g, n_1, \ldots, n_g$ such that

$$[I \quad \Omega]\begin{bmatrix} m_1 \\ \vdots \\ m_g \\ n_1 \\ \vdots \\ n_g \end{bmatrix} = \left[\int_C \omega_j \wedge \psi^1\right]$$

$$= \left[-\sum_k \left(\int_{\alpha_k} \psi\right)\int_C \omega_j \wedge d\eta_k\right]$$

$$\underset{(4.7)}{=} \left[-\int_{\alpha_j} \psi\right].$$

Now replace ψ by

$$\psi - \sum n_j \omega_j.$$

This new meromorphic ψ has integral periods so that the function (4.35) is actually well defined.

4.6 The Jacobi Inversion Theorem

Now we want to do some more work with the mapping

$$\kappa\colon C \to J(C).$$

Actually we should write κ_{p_0} since it depends on a choice of the basepoint p_0. Anyway $J(C)$ is an Abelian complex Lie group, so we have induced analytic mappings

$$\kappa\colon \underbrace{C \times \cdots \times C}_{r \text{ times}} \to J(C).$$

Now this mapping is independent of the ordering of the r-tuple in the domain, and so we have a commutative diagram

$$
\begin{array}{c}
C \times \cdots \times C \quad \overset{\kappa}{\searrow} \\
\downarrow \qquad\qquad \overset{}{\longrightarrow} J(C) \qquad\qquad (4.37)\\
C^{(r)} \quad \overset{\kappa_r}{\nearrow}
\end{array}
$$

where $C^{(r)}$ is the *r-fold symmetric product* of C. $C^{(r)}$ is a smooth r-dimensional complex manifold whose local coordinates are the elementary symmetric functions in the local coordinates on $C \times \cdots \times C$.

Notice that Abel's theorem implies that $\kappa\colon C \to J(C)$ is an embedding if $g \geq 1$, since a curve of genus greater than zero does not admit a meromorphic function with only one pole of order 1. Also the homology class of $\kappa(C)$ in $J(C)$ is

$$\sum_j \alpha_j \times \beta_j, \qquad j = 1, \ldots, g,$$

(where $\times$ is Pontryagin product) since

$$\int_{\kappa(C)} d\xi_j \wedge d\xi_k = \int_C d\xi_j \wedge d\xi_k = 0,$$

$$\int_{\kappa(C)} d\xi_j \wedge d\eta_k = \text{Kronecker } \delta_{jk},$$

$$\int_{\kappa(C)} d\eta_j \wedge d\eta_k = 0.$$

Thus the homology map

$$(\kappa_g)_*\colon H_{2g}(C^{(g)}; \mathbb{Z}) \to H_{2g}(J(C); \mathbb{Z})$$

must be an isomorphism since

$$\kappa_*(C^{(g)}) = \frac{1}{g!} \left(\sum \alpha_j \times \beta_j \right)^g$$

$$= \alpha_1 \times \beta_1 \times \alpha_2 \times \cdots \times \alpha_g \times \beta_g.$$

So by topology, κ_g must be surjective and its sheet number must be one. By the elementary properties of complex analytic morphisms, this can happen only if κ_g is generically injective, that is, if there are open dense sets of $C^{(g)}$ and $J(C)$ which are isomorphic under κ_g. This fact is usually called the *Jacobi inversion theorem*.

After all, for any point $L \subseteq J(C)$, Abel's theorem tells us that

$$\kappa_g^{-1}(L)$$

is a full linear system of divisors of degree g on C (that is, a maximal set of divisors each of which gives the same line bundle L). So $\kappa_g^{-1}(L)$ is natural isomorphic to the projective space

$$\mathbb{P}(H^0(C; \mathcal{O}(L))),$$

where $\mathcal{O}(L)$ is the sheaf of sections of L. Thus the Jacobi inversion theorem says that "most" line bundles of degree g have only a one-dimensional space of sections. By the Riemann–Roch theorem (4.48), this is simply the assertion that, in general, g points on a curve are not contained in the zero set of a nontrivial holomorphic differential.

4.7 Back to Theta Functions

Now let us bring in the concept of theta function in genus g. If δ_j, $\varepsilon_k \in \{0, 1\}$, $j, k = 1, \ldots, g$, define

$$\delta = \begin{bmatrix} \delta_1 \\ \vdots \\ \delta_g \end{bmatrix}, \qquad \varepsilon = \begin{bmatrix} \varepsilon_1 \\ \vdots \\ \varepsilon_g \end{bmatrix},$$

and

$$\theta \begin{bmatrix} \delta \\ \varepsilon \end{bmatrix}(u; \Omega)$$

$$= \sum_{m \in \mathbb{Z}^g} \exp\{\pi i[{}^t(m + \delta/2)\Omega(m + \delta/2) + 2{}^t(m + \delta/2)(u + \varepsilon/2)]\}, \quad (4.38)$$

where $u \in \mathbb{C}^g$. The fact (4.9) that

$$\text{Im } \Omega > 0$$

implies uniform and absolute convergence of this series in a way completely analogous to the one-dimensional case which we saw in Chapter Three. As before, direct computation gives

$$\theta[{}^\delta_\varepsilon](u + E_j; \Omega) = \exp\{-\pi i \delta_j\}\theta[{}^\delta_\varepsilon](u; \Omega),$$
$$\theta[{}^\delta_\varepsilon](u + \Omega_j; \Omega) = \exp\{-\pi i(2u_j + \varepsilon_j + \omega_{jj})\}\theta[{}^\delta_\varepsilon](u; \Omega),$$

$$(4.39)$$

where E_j is the jth column of the identity matrix I and Ω_j is the jth column of Ω. Again we have the task of tracking down the zero set of the functions (4.39). We will do this in a very classical way which goes all the way back to Riemann. In fact, the material contained in the rest of this chapter appears in the last article published by Riemann himself.† Before beginning, we compute

$$
\begin{aligned}
\theta[{}^\delta_\varepsilon](-u; \Omega) &= \sum \exp\{\pi i[{}^t(m + \delta/2)\Omega(m + \delta/2) \\
&\quad + 2{}^t(m + \delta/2)(-u + \varepsilon/2)]\} \\
&= \sum \exp\{\pi i[{}^t(-m - \delta/2)\Omega(-m - \delta/2) \\
&\quad + 2{}^t(-m - \delta/2)(u - \varepsilon/2)]\} \\
&= \sum \exp\{\pi i[{}^t(-m - \delta/2)\Omega(-m - \delta/2) \\
&\quad + 2{}^t(-m - \delta/2)(u + \varepsilon/2) - 2{}^t(-m - \delta/2)\varepsilon]\} \\
&= \exp\{\pi i {}^t\delta \cdot \varepsilon\}\theta[{}^\delta_\varepsilon]
\end{aligned}
$$

Thus the functions (4.38) are even or odd according to whether ${}^t\delta \cdot \varepsilon$ is even or odd. For reasons that will become apparent later, we want to say this in a fancy way. Define

$$[{}^\delta_\varepsilon] = \sum \delta_j \alpha_j + \varepsilon_j \beta_j \in H_1(C; \mathbb{F}_2),$$

where $\mathbb{F}_2$ is the field with two elements and α_j, β_j is the symplectic homology basis that we have been using. Then

$$\mathscr{Q}([{}^\delta_\varepsilon]) = {}^t\delta \cdot \varepsilon \text{ modulo } 2 \qquad (4.40)$$

is a quadratic form on the $\mathbb{F}_2$-vector space $H_1(C; \mathbb{F}_2)$ whose associated bilinear form is the intersection pairing. Now the mapping

$$C \to J(C)$$

lifts to an immersion of the universal covering spaces,

$$\tilde{C} \to H^{1,0}(C)^*.$$

---

† The English translation of the title of Riemann's article is "On the Vanishing of Theta-Functions"; the article appeared in *Borchardt's Journal für reine und angewandte Matematik*, vol. 65, in 1865, the year before Riemann died.

In Figure 4.1 we have simply given a picture of a fundamental domain

$$\hat{C} \subseteq \tilde{C}$$

with respect to the action of the covering group $\pi_1(C, p_0)$. Now using the dual basis

$$\beta_1, \ldots, \beta_g \in H^{1,\,0}(C)^*$$

to

$$\omega_1, \ldots, \omega_g,$$

we get an explicit isomorphism

$$H^{1,\,0}(C)^* \cong \mathbb{C}^g,$$

where the projection of $H_1(C; \mathbb{Z})$ into $H^{1,\,0}(C)^*$ corresponds to the lattice generated by the columns

$$E_j, \Omega_j$$

of the matrix $[I, \Omega]$. Thus

$$J(C) \cong \frac{\mathbb{C}^g}{\sum \mathbb{Z}E_j + \sum \mathbb{Z}\Omega_j}. \tag{4.41}$$

Therefore, by (4.39), the zero set of $\theta[{}_\varepsilon^\delta](u; \Omega)$ gives a well-defined hypersurface in $J(C)$. We will later show that up to translation

$$\left(\text{zero set of } \theta[{}_\varepsilon^\delta](u; \Omega)\right) = \kappa_{(q-1)}(C^{(g-1)}).$$

Before proceeding, we should mention that a more modern version of the theory of theta functions treats them as sections of positive line bundles on abelian varieties. The formulas (4.39) then appear as "patching data" or arise naturally in the cohomology of certain groups associated with abelian varieties. The reader interested in pursuing this direction is referred to Mumford [6].

4.8 The Basic Computation

Now we have shown in genus 1, where $C = J(C)$, that $\int_{\beta_j} \omega = 1$ and $\int_{\alpha_j} \omega = -\tau$ and so

$$\left(\text{zero set of } \theta[{}_\varepsilon^\delta](u; \tau)\right) + [{}_{\varepsilon'}^{\delta'}] = \left(\text{zero set of } \theta[{}_{\varepsilon'+\varepsilon}^{\delta'+\delta}](u; \tau)\right) \tag{4.42}$$

where $[{}_\varepsilon^\delta]$ is considered a point of $J(C)$ according to the rule

$$H_1(C; \mathbb{F}_2) \cong \frac{\tfrac{1}{2}H_1(C; \mathbb{Z})}{H_1(C; \mathbb{Z})} \subseteq \frac{H_1(C; \mathbb{R})}{H_1(C; \mathbb{Z})} = J(C). \tag{4.43}$$

We wish to show that (4.43) is true in general. This fact, together with the assertion at the end of Section 4.7, will be proved with the help of the following essential computation.

Fix a constant vector $e \in \mathbb{C}^g$. Then it makes sense to restrict $\theta[{}^\delta_\varepsilon](u + e; \Omega)$ to $\hat{C}$ and look at its zero set there. A simple computation, completely analogous to the one we did in the case $g = 1$ in Section 3.3 gives that this restriction has g zeros (unless $\theta[{}^\delta_\varepsilon](u + e; \Omega)|_{\hat{C}}$ is identically 0). A deeper computation of the same sort is the following.

Notice that in Figure 4.1 the vector-valued function u (the coordinate function on $\mathbb{C}^g$) satisfies the relations

$$
\begin{aligned}
(u \text{ along } \alpha_j^{-1}) &= (u \text{ along } \alpha_j) + E_j, \\
(u \text{ along } \beta_j^{-1}) &= (u \text{ along } \beta_j) + \Omega_j.
\end{aligned}
\tag{4.44}
$$

So if θ denotes any of the functions (4.38), possibly translated by a constant vector, *restricted* to $\hat{C}$, we compute for e small:

$u[\text{zeros of } \theta(u(\ ) + e)] - u[\text{zeros of } \theta(u(\ ))]$

$$
= \frac{1}{2\pi i} \sum \int_{\alpha_j} u \cdot [d \log \theta(u + e) - d \log \theta(u)]
$$

$$
+ \frac{1}{2\pi i} \sum \int_{\beta_j} u \cdot [d \log \theta(u + e) - d \log \theta(u)]
$$

$$
- \frac{1}{2\pi i} \sum \int_{\alpha_j} (u + E_j)[d \log \theta(u + e + E_j) - d \log \theta(u + E_j)]
$$

$$
- \frac{1}{2\pi i} \sum \int_{\beta_j} (u + \Omega_j)[d \log \theta(u + e + \Omega_j) - d \log \theta(u + \Omega_j)]
$$

$$
\underset{(4.39)}{=} - \frac{1}{2\pi i} \sum \int_{\alpha_j} E_j[d \log \theta(u + e + E_j) - d \log \theta(u + E_j)] \tag{4.45}
$$

$$
- \frac{1}{2\pi i} \sum \int_{\beta_j} \Omega_j[d \log \theta(u + e + \Omega_j) - d \log \theta(u + \Omega_j)]
$$

$$
\underset{(4.39)}{=} - \frac{1}{2\pi i} \sum E_j \Big[\log \theta(u + e + E_j) - \log \theta(u + E_j) \Big]_{u = u^{(j)}}^{u = u^{(j)} - \Omega_j}
$$

$$
- \frac{1}{2\pi i} \sum \Omega_j \Big[\log \theta(u + e + \Omega_j) - \log \theta(u + \Omega_j) \Big]_{u = v^{(j)}}^{u = v^{(j)} + E_j}
$$

$$
\underset{(4.39)}{=} - \frac{1}{2\pi i} \sum E_j \log \left(\frac{\theta(u^{(j)} - \Omega_j + e)}{\theta(u^{(j)} + e)} \div \frac{\theta(u^{(j)} - \Omega_j)}{\theta(u^{(j)})} \right)
$$

$$
\underset{(4.39)}{=} - e.
$$

If e is not small, there is some ambiguity in the choice of the branch of the logarithm in the fourth step in (4.45), but the answer is the same modulo an element of the form

$$\sum m_j E_j + n_j \Omega_j, \qquad m_j, n_j \in \mathbb{Z}.$$

What this means is that if

$$\Theta\begin{bmatrix}\delta\\\varepsilon\end{bmatrix} \subseteq J(C)$$

represents the zero set of $\theta\begin{bmatrix}\delta\\\varepsilon\end{bmatrix}(u; \Omega)$, then

$$\Theta\begin{bmatrix}\delta\\\varepsilon\end{bmatrix} + e$$

represents the zero set of $\theta\begin{bmatrix}\delta\\\varepsilon\end{bmatrix}(u - e; \Omega)$. Identifying C with $\kappa(C) \subseteq J(C)$, we can write

$$\text{(sum of } g \text{ points of } (\Theta\begin{bmatrix}\delta\\\varepsilon\end{bmatrix} + e) \cdot C)$$

$$- (\text{sum of } g \text{ points of } (\Theta\begin{bmatrix}\delta\\\varepsilon\end{bmatrix}) \cdot C) = e. \quad (4.46)$$

Of course, this formula makes no sense if

$$C \subseteq (\Theta\begin{bmatrix}\delta\\\varepsilon\end{bmatrix} + e). \quad (4.47)$$

However, the functions $\theta\begin{bmatrix}\delta\\\varepsilon\end{bmatrix}(u; \Omega)$ are given by Fourier series and so cannot vanish identically on $\mathbb{C}^g$ since not all the Fourier coefficients are 0. Thus (4.47) only holds for those e lying in some proper analytic subvariety of $J(C)$. In a bit we shall see the significance of this subvariety.

First notice the significance of the formula (4.46). It says that the map

$$J(C) \longrightarrow C^{(g)},$$
$$e \longmapsto ((\Theta\begin{bmatrix}\delta\\\varepsilon\end{bmatrix} + e) \cdot C)$$

is (up to translation by a constant), the *inverse* of the map

$$\kappa_g \colon C^{(g)} \to J(C).$$

This is, of course, another way to see the Jacobi inversion theorem.

4.9 Riemann's Theorem

Before proceeding, we should at least state the Riemann–Roch theorem for curves explicitly. A proof is found in Gunning's book on Riemann surfaces [3], p. 111. The theorem states that

$$l(D) - i(D) = \deg D + 1 - g, \quad (4.48)$$

where $D = \sum m_k p_k$, $p_k \in C$, and $\deg D = \sum m_k$ and

$l(D)$ = dimension of vector space of meromorphic functions with at most a pole of order m_k at p_k,

$$= \dim H^0(C; \mathcal{O}(\otimes_k L_{p_k}^{m_k}))$$

and

$i(D)$ = dimension of meromorphic differentials with at least a zero of order m_k at p_k

$$= \dim H^0(C; \mathcal{O}(K \otimes \otimes_k L_{p_k}^{-m_k}))$$

$$\underset{\substack{\text{Serre} \\ \text{duality}}}{=} \dim H^1(C; \mathcal{O}(\otimes L_{p_k}^{m_k})).$$

Now we have seen that for

$$p_1 + \cdots + p_g$$

in an open dense subset of $C^{(g)}$,

$$l(p_1 + \cdots + p_g) = 1.$$

This is just another way of stating the Jacobi inversion theorem. Let us fix

$$\theta(u) = \theta[{}^0_0](u; \Omega).$$

Referring to (4.46), we then define *Riemann's constant* K_{P_0} by the formula

$$\sum \{u(p): p \in [\text{zeros of } \theta(u(\ ) - (e + K_{p_0}))]\} = e \qquad (4.49)$$

in $J(C)$. K_{p_0} clearly depends on the choice of basepoint p_0 used in the embedding

$$\kappa: \quad C \longrightarrow J(C),$$

$$p \longmapsto \left(\int_{p_0}^{\cdot p} \right).$$

Now pick $p_1 + \cdots + p_g$ such that $l(p_1 + \cdots + p_g) = 1$ and such that if $u(p_1) + \cdots + u(p_g) = e$, then

$$\theta(u(\ ) - (e + K_{p_0})) \qquad (4.50)$$

does not vanish identically on C. By formula (4.49) and the fact that no other divisor

$$q_1 + \cdots + q_g$$

has the property that $u(q_1) + \cdots + u(q_g) = e$, the function (4.50) vanishes exactly at $p_1, \ldots, p_g$. We conclude that

$$\theta\left(-u(p_2) - \cdots - u(p_g) - K_{p_0}\right) = 0.$$

Since $p_2 + \cdots + p_g$ can be made to vary over an open dense subset of $C^{(g-1)}$ and since θ is an even function, we have

$$\left(\kappa_{(g-1)}(C^{(g-1)}) + K_{p_0}\right) \subseteq \Theta\begin{bmatrix} 0 \\ 0 \end{bmatrix}.$$

We could have made the same argument (with a different value of K_{p_0}) for any of the $\Theta\begin{bmatrix} \delta \\ \varepsilon \end{bmatrix}$.

Conversely, now, suppose that $\theta(e + K_{p_0}) = 0$. The set of e such that $\theta\left(u(\ ) - e - K_{p_0}\right) \equiv 0$ has codimension ≥ 2 in $J(C)$. Otherwise $\theta\left(u(\ ) - e - K_{p_0}\right)$ has zero set

$$p_1, \ldots, p_g$$

with $u(p_1) + \cdots + u(p_g) = e$. But now if $l(p_1 + \cdots + p_g) = 1$, then the p_k are uniquely determined, so one of them must be the basepoint p_0 since $\theta\left(u(p_0) - e - K_{p_0}\right) = \theta(e + K_{p_0}) = 0$. If $l(p_1 + \cdots + p_g) > 1$, we can always find a divisor

$$p_0 + q_2 + \cdots + q_g$$

such that $u(p_0) + u(q_2) + \cdots + u(q_g) = e$. So in any case

$$e \in \kappa_{(g-1)}(C^{(g-1)}).$$

Thus we have obtained *Riemann's theorem:*

$$\left(\kappa_{(g-1)}(C^{(g-1)}) + K_{p_0}\right) = \Theta\begin{bmatrix} 0 \\ 0 \end{bmatrix}. \tag{4.51}$$

4.10 Linear Systems of Degree g

Next notice that if $l(p_1 + \cdots + p_g) > 1$, then for any pregiven $q \in C$, there exist $q_2, \ldots, q_g$ such that

$$u(q) + u(q_2) + \cdots + u(q_g) = u(p_1) + \cdots + u(p_g) = e.$$

Thus by (4.51)

$$\theta\left(u(\ ) - e - K_{p_0}\right)$$

must vanish at q. So on C

$$\theta\left(u(\ ) - e - K_{p_0}\right) \equiv 0.$$

On the other hand, if

$$\theta(u(\ ) - e - K_{p_0}) \equiv 0,$$

we let $r =$ the maximal integer such that

$$\theta(u(\ ) - (u(p_1) + \cdots + u(p_r)) - e - K_{p_0}) \equiv 0$$

for *all* values of $p_1, \ldots, p_r$. Then for a generic choice of $p_1, \ldots, p_{r+1}$, the function

$$\theta(u(\ ) - (u(p_1) - \cdots - u(p_{r+1})) - e - K_{p_0})$$

has zeros

$$p_1, \ldots, p_{r+1}, q_1, \ldots, q_{g-r-1}.$$

So by (4.49)

$$u(p_1) + \cdots + u(p_{r+1}) + u(q_1) + \cdots + u(q_{g-r-1})$$
$$= u(p_1) + \cdots + u(p_{r+1}) + e$$

so that

$$e \in \kappa_{(g-r-1)}(C^{(g-r-1)}).$$

Since $r \geq 0$ we have shown that

$l(p_1 + \cdots + p_g) = 1$ *if and only if*

$$(\Theta[^0_0] + \sum u(p_k) + K_{p_0}) \not\supseteq \kappa(C), \tag{4.52}$$

in which case the intersection is precisely

$$p_1 + \cdots + p_g.$$

Intuitively it is then clear that we can obtain the entire linear series for $p_1 + \cdots + p_g$ in any case by taking

$$\lim_{e \to 0} \{\kappa(C) \cap [\Theta[^0_0] + \sum u(p_k) + e + K_{p_0}]\}$$

in all possible ways.

4.11 Riemann's Constant

Next let

$\text{Pic}^r(C) =$ space of equivalence classes of line bundles of degree r (Chern class r) on C.

The correspondence

$$\begin{array}{ccc} \text{Pic}^0(C) & \longrightarrow & \text{Pic}^r(C) \\ \{L\} & \longmapsto & \{L \otimes L_{p_0}^r\} \end{array} \tag{4.53}$$

is bijective and so makes $\text{Pic}^r(C)$ a complex torus. Now in $\text{Pic}^{(g-1)}(C)$ we have *intrinsically given* two subvarieties:

(i) $\Theta = \{L \in \text{Pic}^{(g-1)}(C): \dim H^0(C; \mathcal{O}(L)) > 0\}$

   $= $ image in $\text{Pic}^{(g-1)}$ of $C^{(g-1)}$ under the map

$$\kappa^{(g-1)}: C^{(g-1)} \longrightarrow \text{Pic}^{(g-1)}(C),$$

$$(p_1 + \cdots + p_{g-1}) \longmapsto L_{p_1} \otimes \cdots \otimes L_{p_{g-1}};$$

(ii) $\Sigma = \{L \in \text{Pic}^{(g-1)}(C): L \otimes L = \mathcal{K}, \text{ the cotangent bundle of } C\}$.

We call Σ the set of *theta characteristics* of C. Via the mapping (4.53)

$$\kappa_{g-1}(C^{(g-1)}) \leftrightarrow \Theta.$$

Also, by the Riemann–Roch theorem

$$l(p_1 + \cdots + p_{g-1}) = l(K - (p_1 + \cdots + p_{g-1}))$$

whenever K is a *canonical divisor*, that is, a divisor which gives the line bundle $\mathcal{K}$. Thus the mapping

$$u(p_1 + \cdots + p_{g-1}) \mapsto u(K) - u(p_1 + \cdots + p_{g-1})$$

must take $\kappa_{(g-1)}(C^{(g-1)})$ into itself, that is,

$$\kappa_{(g-1)}(C^{(g-1)}) = u(K) - \kappa_{(g-1)}(C^{(g-1)}).$$

But $\theta[{}^0_0](u; \Omega)$ is an even function, so

$$\kappa_{(g-1)}(C^{(g-1)}) + K_{p_0} = -K_{p_0} - \kappa_{(g-1)}(C^{(g-1)}).$$

From these two equations we conclude that

$$u(K) = -2K_{p_0},$$

or in other words $(-K_{p_0})$ corresponds in $\text{Pic}^{(g-1)}(C)$ to a line bundle whose square is the cotangent or *canonical* bundle $\mathcal{K}$, that is, to an element of Σ.

   Next let

$$f(u) = \frac{\theta[{}^\delta_\varepsilon](u; \Omega)}{\theta[{}^0_0](u + I \cdot \varepsilon/2 + \Omega \cdot \delta/2)}.$$

One computes directly from (4.39) that

$$f(u + E_j) = \exp\{\pi i \delta_j\} f(u),$$
$$f(u + \Omega_j) = \exp\{\pi i \delta_j \omega_{jj}\} f(u).$$

Thus the function

$$g(u) = \exp\{-\pi i \Sigma u_j \delta_j\} \cdot f(u)$$

is a well-defined meromorphic function on $J(C)$. Now look at

$$g(u + e)\bigg|_C \tag{4.54}$$

for generic choice of constant e. Either this is a constant function or it has divisor

$$(p_1 + \cdots + p_g) - (q_1 + \cdots + q_g),$$

where $l(q_1 + \cdots + q_g) = 1$. Thus the only possibility is that the function (4.54) is constant for each fixed e and therefore $g(u)$ is constant. Our conclusion is therefore that

$$\Theta\begin{bmatrix} 0 \\ 0 \end{bmatrix} = \Theta\begin{bmatrix} \delta \\ \varepsilon \end{bmatrix} - I \cdot \frac{\varepsilon}{2} - \Omega \cdot \frac{\delta}{2}$$
$$= \Theta\begin{bmatrix} \delta \\ \varepsilon \end{bmatrix} + I \cdot \frac{\varepsilon}{2} + \Omega \cdot \frac{\delta}{2}. \tag{4.55}$$

To put things together we make a diagram

$$
\begin{array}{ccccc}
J(C) & \longrightarrow & J(C) & \xrightarrow{\ (4.53)\ } & \mathrm{Pic}^{(g-1)}(C) \\
u & \longmapsto & u - K_{p_0} & & \\
\Theta\begin{bmatrix} 0 \\ 0 \end{bmatrix} & \longleftarrow & & \longrightarrow & \Theta \\
(?) & \longleftarrow & & \longrightarrow & \Sigma
\end{array}
\tag{4.56}
$$

To see how to fill in the (?), recall that in (4.43) we let

$$\left\{ \begin{bmatrix} \delta \\ \varepsilon \end{bmatrix} \right\}$$

be identified with the set of points of order 2 in $\mathrm{Alb}(C)$. But since

$$2(-K_{P_0} + \begin{bmatrix} \delta \\ \varepsilon \end{bmatrix}) = -2K_{P_0} = u(K)$$

and $(-K_{p_0})$ corresponds to an element of Σ in $\mathrm{Pic}^{(g-1)}(C)$, the same must be true for each element $(-K_{P_0} + \begin{bmatrix} \delta \\ \varepsilon \end{bmatrix})$. Thus via (4.56)

$$\left\{ \begin{bmatrix} \delta \\ \varepsilon \end{bmatrix} \right\} \leftrightarrow \Sigma.$$

Also we should notice that if

$$f(u) = \frac{\theta[\begin{smallmatrix}\delta\\\varepsilon\end{smallmatrix}](u; \Omega)}{\theta[\begin{smallmatrix}0\\0\end{smallmatrix}](u; \Omega)},$$

then $f(u + E_j) = \exp\{\pi i \delta_j\} f(u)$, $f(u + \Omega_j) = \exp\{\pi i \varepsilon_j\} f(u)$. This says, for example, that the line bundle on C given by the divisor

$$(\Theta[\begin{smallmatrix}0\\0\end{smallmatrix}] + \tfrac{1}{2}E_j) - (\Theta[\begin{smallmatrix}0\\0\end{smallmatrix}])$$

is trivial over the set obtained by removing the simple closed path α_j from C and that the bundle

$$(\Theta[\begin{smallmatrix}0\\0\end{smallmatrix}] + \tfrac{1}{2}\Omega_j) - (\Theta[\begin{smallmatrix}0\\0\end{smallmatrix}])$$

becomes trivial if the cycle β_j is removed. So the line bundle of degree 2 in $\text{Pic}^0(C)$ corresponding to $[\begin{smallmatrix}\delta\\\varepsilon\end{smallmatrix}]$ via the Poincaré duality map is simply the one which becomes trivial when a smooth (mod 2) representative of the cycle

$$\sum \delta_j \alpha_j + \varepsilon_j \beta_j$$

is removed.

4.12 Riemann's Singularities Theorem

Now the zero set $\Theta[\begin{smallmatrix}\delta\\\varepsilon\end{smallmatrix}]$ of $\theta[\begin{smallmatrix}\delta\\\varepsilon\end{smallmatrix}](u; \Omega)$ is irreducible since it is given by a translate of

$$\kappa_{g-1}(C^{(g-1)}).$$

It also has multiplicity 1 since we have seen that in general

$$\theta[\begin{smallmatrix}\delta\\\varepsilon\end{smallmatrix}](u + e; \Omega)\Big|_{\hat{C}}$$

has g distinct simple zeros. It turns out that the multiplicity of a point on the theta divisor has a very nice geometric interpretation. This is given in the *Riemann singularities theorem*, which states that if $L \in \text{Pic}^{(g-1)}(C)$, then

$$\text{multiplicity}_L \Theta = \dim H^0(C; \mathcal{O}(L)). \tag{4.57}$$

By then by (4.55) and (4.56):

$$\text{mult}_L \Theta = \text{mult}_k \theta[\begin{smallmatrix}\delta\\\varepsilon\end{smallmatrix}](u; \Omega), \tag{4.58}$$

where

$$k = (L \otimes L_{p_0}^{-(g-1)} + K_{p_0} + I \cdot \varepsilon/2 + \Omega \cdot \delta/2)$$

A corollary of (4.55), (4.58), and the fact that $\theta[^\delta_\varepsilon](u;\Omega)$ is even (odd) if and only if $^t\delta \cdot \varepsilon$ is even (odd) is that if

$$L \in \Sigma \text{ (the set of theta characteristics)},$$

then

$$\mathscr{I}(L) = \dim H^0\big(C;\mathcal{O}(L)\big) \pmod 2$$

is invariant as the curve C is continuously deformed.† It is an easy combinatorial exercise to show that $^t\delta \cdot \varepsilon$ is even exactly $2^{g-1}(2^g + 1)$ times and odd exactly $2^{g-1}(2^g - 1)$ times.

To get some idea why the Riemann singularities theorem is true, let us look at the following. Suppose

$$l(p_1 + \cdots + p_{g-1}) > s.$$

Then given $q_1, \ldots, q_s$, there is a complementary set $q_{s+1}, \ldots, q_{g-1}$ such that

$$\sum_{k=1}^{g-1} u(p_k) = \sum_{k=1}^{g-1} u(q_k) = e.$$

Then for any $p_1, \ldots, p_{g-1}$ and $q_1, \ldots, q_{g-1}$ satisfying this last equation,

$$\theta\left(\sum_{k=1}^{s} u(p_k) - \sum_{k=1}^{s} u(q_k) - e - K_{p_0}\right)$$

$$= \theta\left(-\sum_{k=1}^{s} u(q_k) - \sum_{k=s+1}^{g-1} u(p_k) - K_{p_0}\right) = 0,$$

so that

$$\theta\big(\kappa_s(C^{(s)}) - \kappa_s(C^{(s)}) - e - K_{p_0}\big) \equiv 0. \tag{4.59}$$

Conversely, if (4.59) holds for s but not $s + 1$, then for generic choice of

$$p_1, \ldots, p_s, q_1, \ldots, q_{s+1}$$

the function

$$\theta\left(u(\quad) + \sum_{j=1}^{s} u(p_j) - \sum_{j=1}^{s+1} u(q_j) - e - K_{p_0}\right)$$

has zeros

$$q_1, \ldots, q_{s+1}, q_{s+2}, \ldots, q_g.$$

---

† An algebraic proof of this fact has been given in recent years by David Mumford in "Theta Characteristics of an Algebraic Curve," *Annales scientifiques de l'Ecole Normale Supérieure*, vol. 4, 1971, pp. 181–192.

So by (4.49)

$$\sum_{j=1}^{g} u(q_j) = e + \sum_{j=1}^{s+1} u(q_j) - \sum_{j=1}^{s} u(p_j),$$

that is,

$$\sum_{j=1}^{s} u(p_j) + \sum_{j=s+2}^{q} u(q_j) = e.$$

Since the p_j were chosen arbitrarily, we conclude that

$$l(p_1 + \cdots + p_s + q_{s+2} + \cdots + q_g) > s.$$

Thus

$l(p_1 + \cdots + p_{g-1}) > s$ if and only if

$$\theta\big(\kappa_s(C^{(s)}) - \kappa_s(C^{(s)}) - \sum u(p_k) - K_{p_0}\big) \equiv 0. \quad (4.60)$$

Now if z is a local coordinate at $p \in C$ and $\theta(\kappa(C) - \kappa(C) - e - K_{p_0}) \equiv 0$, then

$$0 = \lim_{q \to p} \frac{\theta\big(u(q) - u(p) - e - K_{p_0}\big)}{z(q) - z(p)} = \sum \frac{\partial \theta}{\partial u_j}(-e - K_{p_0}) \frac{\partial u_j}{\partial z}(p).$$

Repeating the argument at each $p \in C$ we conclude that

$$\sum_j \frac{\partial \theta}{\partial u_j}(-e - K_{p_0})\, \omega_j = 0$$

so that

$$\frac{\partial \theta}{\partial u_j}(-e - K_{p_0}) = 0, \qquad j = 1, \ldots, g.$$

The argument is then repeated to obtain one direction of the Riemann singularities theorem, namely, if

$$\theta\big(\kappa_s(C^{(s)}) - \kappa_s(C^{(s)}) - e - K_{p_0}\big) \equiv 0,$$

then all partial derivatives of θ through order s vanish at $-e - K_{p_0}$. The proof of the other direction is somewhat more difficult, but the general idea is clear.†

Now by the Riemann singularities theorem we can characterize, for example, the set of singular points of Θ, which we denote

$$\Theta_{sg},$$

---

† See J. Lewittes, "Riemann Surfaces and the Theta Function," *Acta Math.* vol. III, 1964, pp. 51–55.

as the set of line bundles of degree $(g - 1)$ with at least two linearly independent sections. To get an intuitive idea about the dimension of this set, we can argue very roughly as follows. The tangent space to the moduli space of curves of genus g at a point C is given by

$$H^1\big(C; \mathcal{O}(\mathcal{T})\big),$$

where $\mathcal{T}$ is the complex tangent bundle of C.[†] By Serre duality, which we saw at the end of Chapter Two, the cotangent space is therefore given by

$$H^0(C; \mathcal{K}^{(2)}),$$

which by the Riemann–Roch theorem has dimension

$$2(2g - 2) + 1 - g = 3g - 3.$$

(Genus 0 and 1 are exceptions to this because curves in these genera have automomorphism groups of dimension > 0.) On the other hand, in how many ways can we make a $(g - 1)$-sheeted covering of $\mathbb{CP}_1$ which is of genus g? Assuming that all the branch points are simple, we compute the number N of branch points via the Euler characteristic formula to get

$$(g - 1) \cdot 2 - N = 2 - 2g,$$
$$N = 4g - 4.$$

Now assume that three of these branch points are 0, 1, ∞ on $\mathbb{CP}_1$; then we are free to move $(4g - 7)$ of them. Since there are only an "∞^{3g-3}" of distinct curves of genus g, we expect an

$$\infty^{(4g-7)-(3g-3)} = \infty^{(g-4)}$$

of coverings to correspond to a fixed generic curve C of genus g. In other words, for a genus C, we expect

$$\dim \Theta_{sg} = (g - 4), \tag{4.61}$$

which indeed turns out to be the case if C is not hyperelliptic.[‡]

[†] See the introduction to K. Kodaira and D. Spencer, "On Deformations of Complex Analytic Structures," *Annals of Mathematics*, vol. 67, 1958, p. 328ff.

[‡] For a proof of this fact, see A. Andreotti and A. Mayer, "On Period Relations for Abelian Integrals on Algebraic Curves," *Ann. Scuola Norm. Sup., Pisa*, vol. 21, 1967, p. 209.

CHAPTER FIVE

Quartics and Quintics

5.1 Topology of Plane Quartics

In this section we will briefly examine some of the more entertaining properties of curves

$$C \subseteq \mathbb{CP}_2 \tag{5.1}$$

defined by equations

$$F(X_0, X_1, X_2) = 0,$$

where F is a homogeneous polynomial of degree 4. We will assume that C is *nonsingular*, that is, the partial derivatives

$$\frac{\partial F}{\partial X_j}, \qquad j = 0, 1, 2,$$

have no common zeros in $\mathbb{CP}_2$.

Any such C is a Riemann surface of genus 3. To see this, form a family

$$t_0 F + t_1 X_2 G = 0 \tag{5.2}$$

of equations of degree 4 by taking G to be a generically chosen polynomial of degree 3.

Corresponding to (5.2) we have a family of curves

$$C_t, \qquad t \in \mathbb{CP}_1, \tag{5.3}$$

where, if $t = (t_0, t_1)$, then C_t is defined by (5.2). Now a small neighborhood of C_t in $\mathbb{CP}_2$ can be made into a disk bundle over C_t as long as C_t is nonsingular. Intersecting normal disks with nearby C_t, one concludes easily that the C_t in (5.3) are diffeomorphic, with the exception of those finite number of values of t for which C_t is singular.

147

One of the singular values of t is, of course,

$$t = (0, 1).$$

As t approaches $(0, 1)$, we can visualize the "moving picture" of C_t by drawing an analogous moving picture in $\mathbb{RP}_2$, the real projective plane (Figure 5.1).

The point of Figure 5.1 is that a nearby C_t is topologically the same as $C_{(0, 1)}$ except that small neighborhoods of the crossing points p_j in $C_{(0, 1)}$, which look like solution sets to

$$x \cdot y = 0,$$

have to be replaced by sets which look like solution sets to

$$x \cdot y = \varepsilon$$

for some constant $\varepsilon \neq 0$. So, in the complex case, we can reconstruct a topological model for the nonsingular C_t by starting with $C_{(0, 1)}$ with its three singular points p_1, p_2, p_3, cutting out a neighborhood of each in $C_{(0, 1)}$, and replacing each neighborhood with a tube. So near each p_j, $C_{(0, 1)}$ used to look like Figure 5.2. But a nearby C_t looks like Figure 5.3.

So what then is a topological model for our nonsingular C_t? We take the set $\{G = 0\}$, which we saw in Chapter Two to be a torus, the set

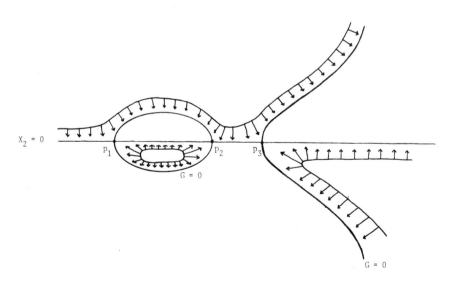

Figure 5.1. Curve with arrows is "moving curve" C_t.

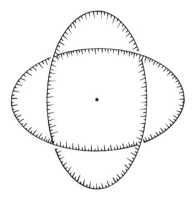

Figure 5.2. Two two-real-dimensional disks meeting transversely at a point in four-real-dimensional space.

$\{X_2 = 0\}$ which is a two-sphere, cut out neighborhoods of p_1, p_2, and p_3 in each, and fill in tubes. Then we get Figure 5.4.

In the same way we see that the genus of an nth-degree curve should be given by the formula

[genus of $(n - 1)$st degree curve] $+ (n - 2)$,

or simply

$$\sum_{k=1}^{n-2} k = \frac{(n-1)(n-2)}{2}.$$

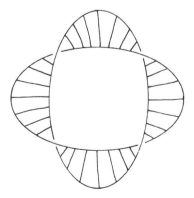

Figure 5.3. A copy of $S^1 \times ((0, 1))$ embedded in four-real-dimensional space.

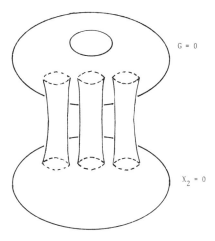

Figure 5.4. Why a fourth-degree curve has genus 3.

5.2 The Twenty-Eight Bitangents

There are two very interesting and related facts about nonsingular quartics.

We recall the mapping

$$\kappa : C \to J(C) \tag{5.4}$$

described in (4.22). Abel's theorem tells us that κ is injective. In fact, the map κ is always an embedding since, by the Riemann–Roch theorem, given any $p \in C$ there is always a holomorphic differential

$$\omega \in H^{1,\,0}(C)$$

such that

$$\omega|_p \neq 0.$$

So we can associate with each $p \in C$ the one-dimensional subspace of the tangent space to $J(C)$ at $\kappa(p)$,

$$A_p \subseteq T_{J(C),\,\kappa(p)},$$

which is tangent to $\kappa(C)$ at $\kappa(p)$. Now the commutative group structure on $J(C)$ induces a unique natural isomorphism of tangent spaces

$$\phi : T_{J(C),\,\kappa(p)} \xrightarrow{\;\cong\;} T_{J(C),\,0}.$$

So we get a well-defined holomorphic map, called the *Gauss map*,

$$\gamma: \quad C \longrightarrow \mathbb{P}(T_{J(C), 0}),$$
$$p \longmapsto \{\phi(A_p)\}, \tag{5.5}$$

where

$\mathbb{P}(\text{vector space } V) = (\text{set of all one-dimensional subspaces of the vector space } V).$

It is a simple exercise to show that the mapping γ is the same as the *canonical mapping*

$$C \longrightarrow \mathbb{P}(H^{1, 0}(C)^*),$$
$$p \longmapsto \{\{\omega \in H^{1, 0}(C): \omega|_p = 0\}\}.$$

So the mapping γ is injective unless there are distinct points p and q on C with the property that every ω which vanishes at p also vanishes at q, that is, referring to the Riemann–Roch theorem in Section 4.9, we get

$$i(p + q) = i(p) = (g - 1).$$

So by the Riemann–Roch theorem

$$l(p + q) - (g - 1) = 2 + 1 - g,$$

and so the curve C must be *hyperelliptic*, that is, there is a meromorphic function f on C which gives a double-branched cover

$$f: C \to \mathbb{CP}_1.$$

Replacing $p + q$ by the divisor $2p$ and making the same argument, we conclude that unless C is hyperelliptic, γ is in fact an embedding of C into a projective space of dimension $(g - 1)$. The degree of $\gamma(C)$ is equal to the degree of the canonical or cotangent bundle to C, which is $(2g - 2)$. In case $g = 3$ we conclude that every nonhyperelliptic curve of genus 3 is (canonically) embedded as a nonsingular quartic in $\mathbb{P}_2$.

Conversely, if C is a nonsingular plane quartic, then the vector space of homogeneous forms of degree 1 on $\mathbb{P}_2$ lies in the space of sections of the line bundle on C whose associated divisor is the intersection of a line L in $\mathbb{P}_2$ with C. Again by the Riemann–Roch theorem

$$3 \leq l(C \cdot L) = i(C \cdot L) + 4 + 1 - 3.$$

so that

$$i(C \cdot L) \geq 1.$$

But since any holomorphic differential has only four zeros, we conclude that in fact these inequalities must actually be equalities and that the hyperplane sections of C must be the *canonical divisors* of C, that is, the zero sets of holomorphic differentials. So up to a linear automorphism of $\mathbb{P}_2$, the inclusion

$$C \subseteq \mathbb{P}_2$$

is simply the Gauss map $\mathscr{G}$. Thus all nonsingular plane quartics are canonically embedded. So there must exist nonhyperelliptic curves of genus 3; in fact, no nonsingular plane quartic is hyperelliptic.

Now we are ready to take up the famous old fact that given a nonsingular plane quartic C, there are exactly 28 lines in $\mathbb{P}_2$ which are tangent to C at each point at which they intersect C. Let's count constants in a very rough way to see why we expect the number 28. Recall that in Section 1.9 we discussed the dual mapping

$$\mathscr{D}: C \to \mathbb{P}_2^* = (\text{set of lines in } \mathbb{P}_2).$$

In the case of the plane quartic the degree of this map is given by

$$\mathscr{D}(C) \cdot (\text{line in } \mathbb{P}_2^*) = C \cdot \left(\text{zero set of } \sum a_j \frac{\partial F}{\partial X_j} \right)$$

$$= 4 \cdot 3.$$

Since we saw that $\mathscr{D}$ is birational onto its image, we expect the genus of $\mathscr{D}(C)$ to be

$$11 \cdot 10/2 = 55.$$

However, $g(C) = 3$ so the only explanation is that $\mathscr{D}(C)$ is singular. In fact, we can analyze the nature of these singularities. First, $\mathscr{D}$ is of maximal rank except at the points at which C intersects the *Hessian curve*

$$\det \begin{bmatrix} \dfrac{\partial^2 F}{\partial X_0^2} & \cdots & \dfrac{\partial^2 F}{\partial X_2 \partial X_0} \\ \vdots & & \\ \dfrac{\partial^2 F}{\partial X_0 \partial X_2} & \cdots & \dfrac{\partial^2 F}{\partial X_2^2} \end{bmatrix} = 0.$$

There are therefore $4 \cdot 6 = 24$ such points, and generically each contributes a simple cusp to $\mathscr{D}(C)$, that is, a singularity which is locally analytically equivalent to

$$y^2 = x^3.$$

Next, suppose that there are two distinct points, p and q, on C such that

$$\mathcal{D}(p) = \mathcal{D}(q).$$

This means that the tangent line L to C at p coincides with that to C at q. Since deg $C = 4$, L must have contact of order exactly 2 with C at each of the two points so that $\mathcal{D}$ is of maximal rank at each. To see that $\mathcal{D}(p) = \mathcal{D}(q)$ is an ordinary crossing point of $\mathcal{D}(C)$, we notice that local coordinates for $\mathbb{P}_2^*$ at L are given by y and z (Figure 5.5). So the local picture at p of a small variation of the tangent line is as shown in Figure 5.6. But in Figure 5.6

$$\lim_{p' \to p} \frac{\text{distance}(p, p'')}{\text{distance}(p, p')} = 0,$$

so that in terms of the local coordinates y, z the Jacobian matrix of $\mathcal{D}$ at p is

$$(0, a)$$

for some $a \neq 0$. Similarly at q the Jacobian matrix of $\mathcal{D}$ is

$$(b, 0).$$

Thus $\mathcal{D}(C)$ has a normal crossing, or ordinary double point, at $\mathcal{D}(p) = \mathcal{D}(q)$. Now we write the exact sheaf sequence

$$0 \to \mathcal{O}_{\mathcal{D}(C)} \to \mathcal{D}_* \mathcal{O}_C \to \mathcal{Q} \to 0, \tag{5.6}$$

where $\mathcal{D}_* \mathcal{O}_C$ on an open set U is the same as $\mathcal{O}_C$ on the open set $\mathcal{D}^{-1}(U)$.

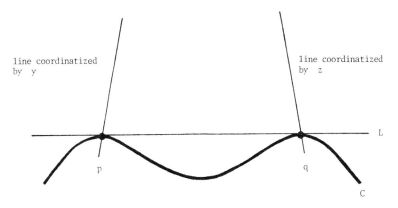

line coordinatized by y

line coordinatized by z

p

q

L

C

Figure 5.5. Local coordinates for the Grassmann variety of lines in $\mathbb{P}_2$ at L.

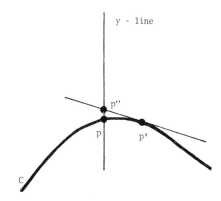

y - line

p''

p

p'

C

Fig. 5.6. How the tangent line to C moves near p.

From our computations it is easy to see that $\mathcal{Q}$ is zero except at the double points and cusp points of $\mathcal{D}(C)$, at each of which points it has as stalk a one-dimensional complex vector space. Now the exactness of the sheaf sequence

$$0 \to \begin{matrix} \mathcal{O}\left(-\mathcal{D}(C)\right) \\ \| \\ \mathcal{O}(-12) \end{matrix} \to \mathcal{O} \to \mathcal{O}_{\mathcal{D}(C)} \to 0$$

shows that the Euler characteristic

$$\chi(\mathcal{O}_{\mathcal{D}(C)}) = \dim H^0\big(\mathcal{D}(C); \mathcal{O}_{\mathcal{D}(C)}\big) - \dim H^1\big(\mathcal{D}(C); \mathcal{O}_{\mathcal{D}(C)}\big)$$

depends only on the degree of $\mathcal{D}(C)$. So $\chi(\mathcal{O}_{\mathcal{D}(C)})$ must equal

$$(1 - 55) = -54$$

since that is what it equals for nonsingular plane curves of degree 12. Returning to (5.6) we therefore conclude that

$$1 - 3 = \chi(\mathcal{O}_C) = \chi(\mathcal{D}_* \mathcal{O}_C) = \chi(\mathcal{Q}) + \chi(\mathcal{O}_{\mathcal{D}(C)})$$

$$= [\text{number of double points of } \mathcal{D}(C)]$$

$$+ [\text{number of cusps of } \mathcal{D}(C)] + (1 - 55)$$

$$= [\text{number of double points of } \mathcal{D}(C)] - 30.$$

So the number of bitangents of C should be 28.

But there is another way to look at the bitangents to a plane quartic. If L is such a bitangent line, then

$$(L \cdot C) = 2p + 2q$$

(where possibly for some special cases $p = q$). Thus $2p + 2q$ is a canonical divisor for C so that the line bundle $L_p \otimes L_q$ associated with the divisor $p + q$ must be a theta characteristic (see Section 4.11). Furthermore, this bundle has at least one section, and if

$$\dim H^0(C; \mathcal{O}(L_p \otimes L_q)) > 1,$$

then C would have to be hyperelliptic, which we know it is not. Thus $L_p \otimes L_q$ is an *odd* theta characteristic. Conversely, any odd theta characteristic has a section. If $(p + q)$ is the zero set of that section then

$$2p + 2q$$

is a canonical divisor and so is the intersection of a line L with C. Thus there is a one-to-one correspondence between bitangents and odd theta characteristics.

But in (4.57) and the argument following it we saw a way to count the number of odd theta characteristics. It was simply the number of solutions $\delta, \varepsilon \in \{0, 1\}^3$ to the equation

$${}^t\delta \cdot \varepsilon \equiv 0 \pmod 2.$$

A direct counting argument shows that this number is

$$2^{(3-1)} \cdot (2^3 - 1) = 28.$$

5.3 Where Are the Hyperelliptic Curves of Genus 3?

Our entire discussion of nonsingular plane quartics in Sections 5.1 and 5.2 is simply a discussion of nonhyperelliptic Riemann surfaces of genus 3. So, we ask ourselves, where do the hyperelliptic curves of genus 3 fit into this discussion? A clue to the answer is provided by examining the Gauss map

$$\mathscr{g} \colon C \to \mathbb{P}(H^{1,0}(C)^*) \tag{5.7}$$

in the case that C is hyperelliptic (see (5.5) and the following discussion). Just as any elliptic curve can be written in the form

$$y^2 = (\text{cubic polynomial in } x)$$

and has holomorphic differential dx/y, one easily checks that any hyperelliptic curve of genus 3 can be written in the form

$$y^2 = (\text{degree-7 polynomial in } x)$$

and has holomorphic differentials

$$\frac{dx}{y},\ \frac{x\,dx}{y},\ \frac{x^2\,dx}{y}.$$

So the Gauss map (5.7) is simply the map

$$
\begin{array}{ccc}
C & \longrightarrow & \mathbb{P}_2 \\
p & \longmapsto & \left(1, x(p), x^2(p)\right),
\end{array}
$$

that is, it realizes C as a double-branched cover of a smooth conic in $\mathbb{P}_2$.

So now suppose we fix a smooth conic A in $\mathbb{CP}_2$ given by

$$G(X_0, X_1, X_2) = 0$$

and take any smooth (nonsingular) quartic C given by

$$F(X_0, X_1, X_2) = 0$$

such that C meets A transversely in eight distinct points. Now form the family of quartics $\{C_t\}$ given by

$$tF + G^2 = 0 \tag{5.8}$$

for $t \in \Delta_\varepsilon$, the open disk of radius ε about 0 in $\mathbb{C}$. We picture this family in Figure 5.7.

We next form

$$\{(s, x) \in \Delta_{\sqrt{\varepsilon}} \times \mathbb{CP}_2 : s^2 F(x) + G(x)^2 = 0\},$$

which we can picture as in Figure 5.8.

Finally, we pull apart or separate Figure 5.8 along the locus where it crosses itself. This last can indeed be done algebraically (by a process called normalization), and the resulting picture is as in Figure 5.9.

Now it can be shown that the family of curves $\{C_s\}$ with $\tilde{C}_0$ put in at $s = 0$ is in fact a smooth family, implying that $\tilde{C}_0$ is in fact diffeomorphic to C_s, $s \neq 0$. Our conclusion is that as t approaches zero, then C_t approaches

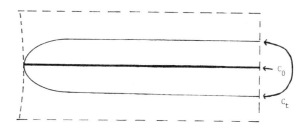

Figure 5.7. A family of quartics lying down doubly over a conic.

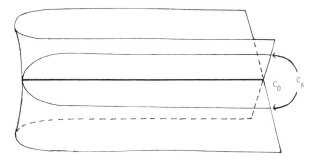

Figure 5.8. Putting two isomorphic curves, C_s and C_{-s}, over C_t for $t \neq 0$.

the hyperelliptic curve of genus 3 which is a double cover of the conic C_0 branched at the eight points

$$C_t \cap C_0.$$

Our goal here is not to be precise but rather to visualize geometrically the behavior of curves of genus 3. In this spirit we can say that the hyperelliptic curves of genus 3 are all concentrated in the various directions of arriving at the double conic G^2 by families (5.8).

It is an interesting little exercise to see what happens to the 28 bitangents of C_t as $t \to 0$. The answer, which can be guessed either from geometric or numerological considerations, is that the 28 bitangents go to the

$$\binom{8}{2} = 28$$

bisecants of the set $C_t \cap C_0$, the set of branch points of

$$\tilde{C}_0 \to C_0. \tag{5.9}$$

That is, the odd theta characteristics of the hyperelliptic curve (5.9) are simply the line bundles

$$L_p \otimes L_q$$

on $\tilde{C}_0$ where p and q are distinct ramification points of (5.9).

Figure 5.9. Construction of $\tilde{C}_0$, a branched double cover of C_0.

5.4 Quintics

We will end this chapter with some observations about plane curves of degree 5. Here a new phenomenon occurs. The "general" curve of genus 6 is not a plane quintic even though nonsingular plane quintics give a nice class of curves of genus 6. We can see this by a constant count. As we remarked in Section 4.12, the family of curves of genus g has $3g-3$ parameters, whereas plane quintics can have no more than

$$21 - 9 = 12$$

parameters, since the vector space of homogeneous forms of degree 5 is $\binom{5+2}{2} = 21$, while the dimension of the group $\mathrm{GL}(3; \mathbb{C})$ which acts on $\mathbb{P}_2$ and so on plane quintics is nine.

One nice property of quintics is that their canonical bundle is obtained by restricting the bundle $\mathcal{O}(2)$ on $\mathbb{C}\mathbb{P}_2$ to the curve. Thus the set of canonical divisors associated with a quintic C is obtained by intersecting C in all possible ways with conics in $\mathbb{C}\mathbb{P}_2$. A quick way to see this is to notice that by the Riemann–Roch theorem, if A is a conic,

$$i(C \cdot A) \geq 6 - 10 - 1 + 6 = 1.$$

This is the same proof we used to show that for a quartic, $\mathcal{O}(1)$ restricted to give the canonical bundle.

This means that quintics C have a distinguished theta characteristic, namely,

$$\mathcal{O}(1)|_C.$$

From the exact sheaf sequence

$$0 \to \mathcal{O}(-4) \to \mathcal{O}(1) \to \mathcal{O}(1)|_C \to 0$$

and the Kodaira vanishing theorem (see Griffiths and Harris [1], p. 155), it is easy to see that $\mathcal{O}(1)|_C$ is an odd theta characteristic; in fact,

$$\dim H^0(C; \mathcal{O}(1)|_C) = 3.$$

The Riemann singularities theorem tells us that this happens exactly when the theta divisor of $J(C)$ has a triple point.

Now plane quintics have another special property; they are closely related to curves of genus 5. This fact was first noticed by the Russian mathematician A. N. Tjurin. We shall omit any discussion of the pathology of special cases of the connection we are about to make. Our assertions will refer to the connection between the *general* plane quintic and the *general* curve of genus 5 (notice that both form 12-dimensional families).[†]

Our first assertion then, true only "generically," is that a curve of genus 5 is the complete intersection of three quadrics

$$Q_0, Q_1, Q_2$$

in $\mathbb{CP}_4$, that is, it is the base (or fixed) locus of a family

$$\lambda Q_0 + \mu Q_1 + \nu Q_2 = 0 \tag{5.10}$$

of homogeneous forms of degree 2 on $\mathbb{CP}_4$. Now each Q_j is given by a symmetric 5×5 matrix $(q_j^{\alpha\beta})$, and the set of those quadrics (5.10) which are *singular* is given by the fifth-degree equation

$$\det \left(\lambda q_0^{\alpha\beta} + \mu q_1^{\alpha\beta} + \nu q_2^{\alpha\beta}\right) = 0 \tag{5.11}$$

in the $\mathbb{CP}_2$ with homogeneous coordinates λ, μ, ν.

But the connection between curves of degree 5 and curves of genus 5 is much richer than this. The curve of genus 5, which we shall call

$$B,$$

which sits in $\mathbb{CP}_4$ as the base locus of the family (5.10), is canonically embedded (same proof as for quartics). So as before, we think of B as lying in

$$\mathbb{P}\left(T_{J(B),\, 0}\right)$$

and of our family of quadrics (5.10) as lying there also. This family is simply the family of all quadrics containing the canonically embedded curve of genus 5. There is another way to produce quadrics containing B in $\mathbb{P}(T_{J(B),\, 0})$, namely, by studying the theta function

$$\theta(u) = \theta\begin{bmatrix} 0 \\ 0 \end{bmatrix}(u; \Omega)$$

on $J(B)$. Suppose $u_0 \in J(B)$ is such that

$$\theta(u_0 + K_{p_0}) = 0, \qquad \frac{\partial \theta}{\partial u_j}(u_0 + K_{p_0}) = 0,$$

where $u = (u_1, \ldots, u_5)$. We then know that u_0 is a singular point of the theta divisor $u(B^{(4)})$. By the Riemann singularities theorem in Section 4.12

[†] The point of view employed in what follows was introduced in the fundamental paper by A. Andreotti and A. Mayer, "On Period Relations for Abelian Integrals on Algebraic Curves," *Ann. Scuola Norm. Sup., Pisa*, vol. 21, 1967, pp. 189–238.

and (4.60),

$$\theta\big(u(p) - u(q) - u_0 - K_{po}\big) = 0$$

for all $p, q \in B$. Differentiating at $p = p_1$ we obtain

$$\sum \frac{\partial \theta}{\partial u_j}\big(u(p_1) - u(q) - u_0 - K_{po}\big)\omega_j(p_1) = 0,$$

where ω_j is the jth element of the basis of $H^{1,\,0}(B)$. Next differentiate with respect to q at $q = p_1$ to obtain that for all $p \in B$

$$\sum \frac{\partial^2 \theta}{\partial u_j \, \partial u_k}\big(-u_0 - K_{po}\big)\omega_j(p)\omega_k(p) = 0.$$

Since θ is even, we can replace $\big(-u_0 - K_{po}\big)$ by $\big(u_0 + K_{po}\big)$ and conclude that the tangent cone

$$\sum_{i \le j,\, k \le 5} \frac{\partial^2 \theta}{\partial u_j \, \partial u_k}\big(u_0 + K_{po}\big)X_j X_k = 0 \tag{5.12}$$

to $u(B^{(4)})$ at u_0 contains B. Thus each point of the singular locus

$$\Theta_{sg} \subseteq u(B^{(4)}) \tag{5.13}$$

has, as its tangent cone, a quadric of the family (5.10). By (4.61), $\dim \Theta_{sg} = 1$. In fact, it can be shown that (again "generically") Θ_{sg} is an irreducible smooth curve.

Also, the cone (5.12) is singular. This is simply because u_0 is not an isolated singular point of $u(B^{(4)})$ as it would be if (5.12) were nondegenerate. In this way we see that there is a mapping

$$\begin{aligned} \pi: \quad &\Theta_{sg} \xrightarrow{\hspace{2cm}} C, \\ &u_0 \longmapsto \text{quadric (5.12)}, \end{aligned} \tag{5.14}$$

where C is the plane quintic defined in (5.11). This map assigns to each u_0 the value of (λ, μ, ν) which gives the tangent cone to $u(B^{(4)})$ at that point. Again generically, it can be shown that (5.14) is a smooth unbranched double covering. That it is at least a double covering is clear since u_0 and $\big(-u_0 - 2K_{po}\big)$ clearly go to the same point.

CHAPTER SIX

The Schottky Relation

6.1 Prym Varieties

In this last chapter we will discuss the most famous classical result in the direction of answering the question, Which symmetric matrices Ω with positive definite imaginary parts arise from curves in the manner presented in Chapter Four? This is the so-called Schottky problem, and in approaching it, we will closely follow the work of H. Farkas.[†]

At the end of Chapter Five we saw a special case of the following phenomenon. Suppose

$$\pi: \tilde{C} \to C \tag{6.1}$$

is an unbranched irreducible double covering of a curve C of genus $(g + 1)$. If we fix a basepoint $p_0 \in C$, then (6.1) corresponds to a subgroup of index 2 in the fundamental group

$$\pi_1(C, p_0).$$

Since this subgroup contains the commutator subgroup of $\pi_1(C, p_0)$, as well as the subgroup of squares in $\pi_1(C, p_0)$, we conclude that the set of double coverings (6.1) of a fixed C is in one-to-one correspondence with the set of subspaces of index 2 in

$$H_1(C; \mathbb{F}_2). \tag{6.2}$$

Since the intersection pairing is nondegenerate on (6.2), each nonzero element of (6.2) corresponds uniquely to a subspace of index 2, namely its

---

[†] "On the Schottky Relation and Its Generalization to Arbitrary Genus," *Annals of Mathematics*, vol. 92, 1970, pp. 56–81.

perpendicular subspace with respect to the intersection pairing. Thus the set of double covers (6.1) is in a natural one-to-one correspondence with the set of nonzero elements of $H_1(C; \mathbb{F}_2)$. If γ is a simple closed curve representing a nonzero element of $H_1(C, \mathbb{F}_2)$, then the covering (6.1) corresponds to γ if and only if $\pi^{-1}(\gamma)$ disconnects $\tilde{C}$.

Now there is a natural involution

$$\iota: \tilde{C} \to \tilde{C} \tag{6.3}$$

over C, and so a natural involution

$$\iota^*: H^{1,0}(\tilde{C}) \to H^{1,0}(\tilde{C}) \tag{6.4}$$

with eigenvalues ± 1. Also let us pick a basis

$$\alpha_0, \alpha_1, \ldots, \alpha_g, \beta_0, \beta_1, \ldots, \beta_g$$

for $H_1(C; \mathbb{Z})$ such that β_0 (reduced mod 2) corresponds to the covering (6.1) and such that the basis is symplectic in the sense of Section 4.1. Then we can build a symplectic basis for $H_1(\tilde{C}; \mathbb{Z})$ as shown in Figure 6.1. For our symplectic basis for $\tilde{C}$,

$$\pi_*(\beta) = \beta_0, \qquad \pi_*(\alpha) = 2\alpha_0,$$
$$\pi_*(\beta_j') = \pi_*(\beta_j'') = \beta_j,$$
$$\pi_*(\alpha_j') = \pi_*(\alpha_j'') = \alpha_j.$$

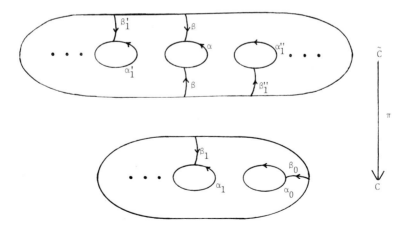

Figure 6.1. A symplectic basis for $H_1(\tilde{C}; \mathbb{Z})$.

Then the cycles

$$\alpha'_j - \alpha''_j, \qquad \beta'_j - \beta''_j, \qquad j = 1, \ldots, g,$$

form a basis for the skew-symmetric part of $H_1(\tilde{C}; \mathbb{Z})$ with respect to the involution ι_*, and the intersection pairing restricted to the skew-symmetric subspace has matrix

$$2\begin{bmatrix} 0 & I \\ -I & 0 \end{bmatrix}.$$

As in (4.7) we choose a basis

$$\psi_1, \ldots, \psi_g$$

for $H^{1,0}(\tilde{C})^-$, the (-1)-eigenspace of $H^{1,0}(\tilde{C})$ with respect to the involution ι^*, such that

$$\int_{\beta'_j - \beta''_j} \psi_k = (\text{Kronecker } \delta_{jk}).$$

We know that $H^{1,0}(\tilde{C})^-$ has dimension g because $\tilde{C}$ has genus $2g + 1$ and

$$H^{1,0}(\tilde{C})^+ \cong H^{1,0}(C).$$

Next we let

$$\Upsilon$$

denote the matrix

$$-\int_{\alpha'_j - \alpha''_j} \psi_k.$$

The matrix Υ takes the place of the matrix Ω in Section 4.2, and the differentials ψ_k, called *Prym differentials*, take the place of the ω_k's of Chapter Four. So just as in Section 4.2, we obtain the Riemann relations

$$\Upsilon = {}^t\Upsilon$$

and

$$\text{Im } \Upsilon > 0.$$

So we can build a complex torus

$$P(\pi) = \frac{\left(H^{1,0}(\tilde{C})^-\right)^*}{H_1(\tilde{C}; \mathbb{Z})^-} \simeq \frac{\mathbb{C}^g}{\sum \mathbb{Z}E_j + \sum \mathbb{Z}\Upsilon_j}, \qquad (6.5)$$

called the *Prym variety* of π, in a manner analogous to the construction of $\text{Pic}^0(C) = J(C)$ in Chapter Four. Also, since the Riemann relations were all that were needed to construct theta functions, we have them in this setting.

Since we will want to compare these with the theta functions on $J(C)$, we will use the letter η rather than θ to denote these functions. Thus

$$\eta[^{\delta}_{\varepsilon}](v;\Upsilon) = \sum_{\mathbb{Z}^g} \exp\{\pi i[^t(m + \tfrac{1}{2}\delta)\Upsilon(m + \tfrac{1}{2}\delta) + 2^t(m + \tfrac{1}{2}\delta)(v + \tfrac{1}{2}\varepsilon)]\}. \quad (6.6)$$

We ought to mention that if

$$\pi\colon \Theta_{s_y} \to C$$

is the double covering of the plane quintic constructed in Section 5.4 in connection with the curve B of genus 5, then

$$P(\pi) = J(B).\dagger$$

Now given any symmetric $g \times g$ matrix A such that

$$\mathrm{Im}\; A > 0,$$

we have seen that we have associated the whole apparatus of complex torus

$$\frac{\mathbb{C}^g}{\sum \mathbb{Z}E_j + \sum \mathbb{Z}A_j},$$

theta functions, etc. We call such a matrix A a *period matrix*. Counting constants, we see that if $g \geq 4$, not every such A will be the "period matrix" Ω of a curve as in Chapter Four. In particular, for $g = 4$ the set of period matrices is a ten-dimensional analytic set, while the set of curves of genus 4 has only

$$3g - 3 = 9$$

dimensions. Thus there must be a nontrivial (analytic) relation on the entries of A that will always be satisfied if A comes from a curve. One such relation, called the *Schottky relation*, is the object of the rest of this chapter.

6.2 Riemann's Theta Relation

To obtain the Schottky relation, we need two main ingredients. The first is the Prym variety construction, which we have already seen. The second is a beautiful trick using the character formula for finite groups, which results in the most far-reaching of the many identities among theta

---

† For more information about this topic as well as recent advances in the theory of Prym varieties, see A. Beauville, "Prym Varieties and the Schottky Problem," *Invent. Math.*, vol. 41, 1977, pp. 149–196.

functions, Riemann's theta relation. We have already derived this relation in the case $g = 1$ in Section 3.14 by different means.

To begin, we will rederive Riemann's theta relation in the case $g = 1$. Let

$$e_1, \ldots, e_4$$

denote the four standard basis vectors in $\mathbb{R}^4$. We put the standard scalar product on $\mathbb{R}^4$ so that these vectors form an orthonormal basis. We next consider two lattices in $\mathbb{R}^4$:

$$L_1 = \mathbb{Z}e_1 \oplus \mathbb{Z}e_2 + \mathbb{Z}e_3 + \mathbb{Z}e_4,$$
$$L_2 = [\text{submodule of } \tfrac{1}{2} \cdot L_1 \text{ generated by the vectors } e_j + e_k, \ 1 \leq j,$$
$$k \leq 4, \text{ and the vector } \tfrac{1}{2}(e_1 + e_2 + e_3 + e_4)].$$

Now $(L_1 \cap L_2)$ is of index 2 in both L_1 and L_2, and so the scalar product is unimodular on L_2, that is, the matrix giving the scalar product for any given basis of L_2 has determinant 1. As scalar product spaces,

$$L_1 \cong L_2.$$

In fact, an explicit orthonormal basis of L_2 is given by

$$\begin{aligned}
f_1 &= \tfrac{1}{2}(e_1 + e_2 + e_3 + e_4), \\
f_2 &= \tfrac{1}{2}(e_1 + e_2 - e_3 - e_4), \\
f_3 &= \tfrac{1}{2}(e_1 - e_2 + e_3 - e_4), \\
f_4 &= \tfrac{1}{2}(e_1 - e_2 - e_3 + e_4).
\end{aligned} \tag{6.7}$$

Notice that if e and f are interchanged, formula (6.7) continues to hold. So the matrix M which gives the isomorphism of L_1 and L_2 satisfies

$$M^2 = (\text{identity}). \tag{6.8}$$

Finally we let

$$L = L_1 + L_2 \tag{6.9}$$

so that $[L:L_2] = 2$ and the extension is generated by any $e_j, j = 1, \ldots, 4$, and $[L:L_1] = 2$ with this latter extension generated by $\tfrac{1}{2}(e_1 + \cdots + e_4)$.

Next suppose that the complex number A satisfies the Riemann relation Im $A > 0$, and we write the product of four theta functions

$$\theta[\begin{smallmatrix} g_1 \\ h_1 \end{smallmatrix}](u_1; A)\theta[\begin{smallmatrix} g_2 \\ h_2 \end{smallmatrix}](u_2; A)\theta[\begin{smallmatrix} g_3 \\ h_3 \end{smallmatrix}](u_3; A)\theta[\begin{smallmatrix} g_4 \\ h_4 \end{smallmatrix}](u_4; A). \tag{6.10}$$

Here the g_j's and h_j's can be arbitrary elements of $\mathbb{R}^g$—we need not restrict ourselves to g-tuples whose entries are all zeros and ones. The same definition of theta function as in (4.38) works in this slightly more general setting. Suppose now we multiply out the Fourier expansions of the theta functions in (6.10). We obtain

$$\exp\{2\pi i(\qquad)\},$$

where the expression which goes in the parentheses is

$$\sum_{j=1}^{4} \tfrac{1}{2}(m_j + \tfrac{1}{2}g_j)^2 A + (m_j + \tfrac{1}{2}g_j)(u_j + \tfrac{1}{2}h_j)$$

$$= \tfrac{1}{2}{}^t(m + \tfrac{1}{2}g)\mathscr{A}(m + \tfrac{1}{2}g) + {}^t(m + \tfrac{1}{2}g)(u + \tfrac{1}{2}h)$$

where

$$m = \begin{bmatrix} m_1 \\ \vdots \\ m_4 \end{bmatrix}, \qquad \text{etc.,}$$

and

$$\mathscr{A} = \begin{bmatrix} A & & & \\ & A & & \\ & & A & \\ & & & A \end{bmatrix}.$$

So we can rewrite (6.10) as

$$\sum_{m \in L_1} \exp\{2\pi i[\tfrac{1}{2}{}^t(m + \tfrac{1}{2}g)\mathscr{A}(m + \tfrac{1}{2}g) + {}^t(m + \tfrac{1}{2}g)(u + \tfrac{1}{2}h)]\}$$

$$= \frac{1}{2}\Bigg\{ \sum_{m \in L} 1 \cdot \exp\{2\pi i[\tfrac{1}{2}{}^t(m + \tfrac{1}{2}g)\mathscr{A}(m + \tfrac{1}{2}g) + {}^t(m + \tfrac{1}{2}g)(u + \tfrac{1}{2}h)]\}$$

$$+ \sum_{m \in L} \exp\{2\pi i(m \cdot e_1)\}\exp\{2\pi i[\tfrac{1}{2}{}^t(m + \tfrac{1}{2}g)\mathscr{A}(m + \tfrac{1}{2}g)$$

$$+ {}^t(m + \tfrac{1}{2}g)(u + \tfrac{1}{2}h)]\Bigg\}, \tag{6.11}$$

where $(m \cdot e_1)$ refers to the standard scalar product mentioned earlier.

Next we break down each of the two sums in (6.11) into two parts according to whether the index of the summand lies in the subgroup L_2 of L. We obtain

$$\frac{1}{2}\sum_{L_2}\left\{ \begin{array}{l} \exp\{2\pi i[\tfrac{1}{2}{}^t(m + \tfrac{1}{2}g)\mathscr{A}(m + \tfrac{1}{2}g) + {}^t(m + \tfrac{1}{2}g)(u + \tfrac{1}{2}h)]\} \\[4pt] + \exp\{2\pi i[\tfrac{1}{2}{}^t(m + e_1 + \tfrac{1}{2}g)\mathscr{A}(m + e_1 + \tfrac{1}{2}g) \\[4pt] + {}^t(m + e_1 + \tfrac{1}{2}g)(u + \tfrac{1}{2}h)]\} \\[4pt] + \exp\{2\pi i(-\tfrac{1}{2}g \cdot e_1)\}\exp\{2\pi i[\tfrac{1}{2}{}^t(m + \tfrac{1}{2}g)\mathscr{A}(m + \tfrac{1}{2}g) \\[4pt] + {}^t(m + \tfrac{1}{2}g)(u + \tfrac{1}{2}h + e_1)]\} \\[4pt] + \exp\{2\pi i(-\tfrac{1}{2}g \cdot e_1)\}\exp\{2\pi i[\tfrac{1}{2}{}^t(m + \tfrac{1}{2}g + e_1)\mathscr{A}(m + \tfrac{1}{2}g + e_1) \\[4pt] + {}^t(m + \tfrac{1}{2}g + e_1)(u + \tfrac{1}{2}h + e_1)]\}. \end{array} \right.$$

Now we use that M is an isometry on $\mathbb{R}^4$ to get the final form of (6.10):

$$\frac{1}{2}\sum_{L_1}\left\{\begin{array}{l}\exp\{2\pi i[\frac{1}{2}{}^t(m+\frac{1}{2}Mg)\mathscr{A}(m+\frac{1}{2}Mg)+{}^t(m+\frac{1}{2}Mg)(Mu+\frac{1}{2}Mh)]\}\\[4pt]\quad+\exp\{2\pi i[\frac{1}{2}{}^t(m+\frac{1}{2}Mg+(\frac{1}{2})))\mathscr{A}(m+\frac{1}{2}Mg+(\frac{1}{2}))\\[4pt]\quad+{}^t(m+\frac{1}{2}Mg+(\frac{1}{2}))(Mu+\frac{1}{2}Mh)]\}\\[4pt]\quad+\exp\{-\pi ig_1\}\exp\{2\pi i[\frac{1}{2}{}^t(m+\frac{1}{2}Mg)\mathscr{A}(m+\frac{1}{2}Mg)\\[4pt]\quad+{}^t(m+\frac{1}{2}Mg)(Mu+\frac{1}{2}Mh+(\frac{1}{2}))]\}\\[4pt]\quad+\exp\{-\pi ig_1\}\exp\{2\pi i[\frac{1}{2}{}^t(m+\frac{1}{2}Mg+(\frac{1}{2}))\mathscr{A}(m+\frac{1}{2}Mg+(\frac{1}{2}))\\[4pt]\quad+{}^t(m+\frac{1}{2}Mg+(\frac{1}{2}))(Mu+\frac{1}{2}Mh+(\frac{1}{2}))]\}\end{array}\right.$$

where $(\frac{1}{2})=(\frac{1}{2},\frac{1}{2},\frac{1}{2},\frac{1}{2})$. Thus

$$\prod_{j=1}^{4}\theta[{}^{g_j}_{h_j}](u_j;A)=\frac{1}{2}\sum_{(^{x'}_{x''})}\exp\{-\pi i[\alpha''\cdot g_1]\}\prod_{j=1}^{4}\theta[{}^{g_j'+x'}_{h_j'+x''}](u_j';A),$$

where $u'=Mu$, $g'=Mg$, and $h'=Mh$ and the index $(^{x'}_{x''})$ runs over all elements of $\mathbb{Z}\times\mathbb{Z}$ whose entries contain only zeros and ones. A special case of this last formula is the identity

$$\theta[^0_0]^4=\tfrac{1}{2}(\theta[^0_0]^4+\theta[^0_1]^4+\theta[^1_0]^4),$$

which we already saw at the end of Chapter Three.

The proof of the corresponding general formula in dimension g differs only notationally from the one-dimensional proof. The matrix M is replaced by a $4g\times4g$ matrix which has a $g\times g$ identity matrix in each place that M had a one. The g_j's, h_j's, u_j's are all now vectors, and in step (6.11) we have 2^g sums corresponding to the 2^g characters of $\mathbb{Z}^g$ with values in the group $\{\pm1\}$. So the final, general form of *Riemann's theta relation* is

$$\prod_{j=1}^{4}\theta[{}^{g_j}_{h_j}](u_j;A)=\frac{1}{2^g}\sum_{(^{x'}_{x''})}\exp\{-\pi i[\alpha''\cdot g_1]\}\prod_{j=1}^{4}\theta[{}^{g_j'+x'}_{h_j'+x''}](u_j';A)\quad(6.12)$$

where u', g', h' are as before and the index $(^{x'}_{x''})$ runs over all elements of $\mathbb{Z}^g\times\mathbb{Z}^g$ whose entries contain only zeros and ones.

6.3 Products of Pairs of Theta Functions

In order to get at the Schottky relation we need another preliminary theta relation which is obtained by a simple variation of the argument used to obtain Riemann's theta relation. Again we start with the case $g=1$. This time we use the matrix

$$M = \begin{bmatrix} 1 & 1 \\ 1 & -1 \end{bmatrix}$$

so that

$$M^2 = 2(\text{identity}).$$

We let

$$L_1 = \mathbb{Z}e_1 + \mathbb{Z}e_2$$

be the standard lattice in $\mathbb{R}^2$ and let

$$L_2 = M^{-1}L_1.$$

Notice that $L_2 \supseteq L_1$ and, for example, $\frac{1}{2}(e_1 + e_2)$ generates L_2/L_1. Also $(\frac{1}{2}(e_1 + e_2) \cdot \frac{1}{2}(e_1 + e_2)) = \frac{1}{2}$ so that the standard scalar product takes half-integral values on L_2.

Now, as in Section 6.2, we write

$$\theta\begin{bmatrix} g_1 \\ h_1 \end{bmatrix}(u; A)\theta\begin{bmatrix} g_2 \\ h_2 \end{bmatrix}(u_2; A)$$

$$= \sum_{m \in L_1} \exp\{2\pi i [\tfrac{1}{2}{}^t(m + \tfrac{1}{2}g).\mathscr{A}(m + \tfrac{1}{2}g) + {}^t(m + \tfrac{1}{2}g)(u + \tfrac{1}{2}h)]\}$$

$$= \sum_{m \in L_2} \exp\{2\pi i [\tfrac{1}{2}{}^t(m + \tfrac{1}{2}g')2\mathscr{A}(m + \tfrac{1}{2}g') + {}^t(m + \tfrac{1}{2}g')(u' + \tfrac{1}{2}h')]\}$$

$$\tag{6.13}$$

where $g' = M^{-1}g$, $h' = Mh$, and $u' = Mu$. But now we split the final sum in (6.13) into two sums, according to the cosets of L_1 in L_2, and we obtain

$$\theta\begin{bmatrix} g_1 \\ h_1 \end{bmatrix}(u; A)\theta\begin{bmatrix} g_2 \\ h_2 \end{bmatrix}(u_2; A) = \sum_{\alpha'} \theta\begin{bmatrix} g_1' + \alpha \\ h_1 \end{bmatrix}(u_1'; 2A)\theta\begin{bmatrix} g_2' + \alpha \\ h_2' \end{bmatrix}(u_2'; 2A), \tag{6.14}$$

where α' ranges over 0 and 1. In fact, as in Section 6.2, the same formula (6.14) holds in all dimensions. The only difference is that α' ranges over all elements of $\mathbb{Z}^g$ whose entries consist only of zeros and ones.

6.4 A Proportionality Theorem Relating Jacobians and Pryms

We are now ready to bring together the ideas of the first three sections of this chapter. Let

$$\pi \colon \tilde{C} \to C$$

be as in (6.1). We then have the period matrices

$$\Omega = \begin{bmatrix} \omega_{00} & \omega_{0g} \\ \vdots & \vdots \\ \omega_{g0} & \omega_{gg} \end{bmatrix}$$

for $J(C)$ and

$$\Upsilon = \begin{bmatrix} \tau_{11} & \tau_{1g} \\ \vdots & \vdots \\ \tau_{g1} & \tau_{gg} \end{bmatrix}$$

for $P(\pi)$. It is then easy to calculate the period matrix for $J(\tilde{C})$ with respect to the homology basis given in Section 6.1. It is

$$\tilde{\Omega} = \begin{bmatrix} 2\omega_{00} & \omega_{01} & \cdots & \omega_{0g} & \omega_{01} & \cdots & \omega_{0g} \\ \omega_{10} & \frac{1}{2}(\omega_{11}+\tau_{11}) & \cdots & \frac{1}{2}(\omega_{1g}+\tau_{1g}) & \frac{1}{2}(\omega_{11}-\tau_{11}) & \cdots & \frac{1}{2}(\omega_{1g}-\tau_{1g}) \\ \vdots & \vdots & & \vdots & \vdots & & \vdots \\ \omega_{g0} & \frac{1}{2}(\omega_{g1}+\tau_{g1}) & \cdots & \frac{1}{2}(\omega_{gg}+\tau_{gg}) & \frac{1}{2}(\omega_{g1}-\tau_{g1}) & \cdots & \frac{1}{2}(\omega_{gg}-\tau_{gg}) \\ \omega_{10} & \frac{1}{2}(\omega_{11}-\tau_{11}) & \cdots & \frac{1}{2}(\omega_{1g}-\tau_{1g}) & \frac{1}{2}(\omega_{11}+\tau_{11}) & \cdots & \frac{1}{2}(\omega_{1g}+\tau_{1g}) \\ \vdots & \vdots & & \vdots & \vdots & & \vdots \\ \omega_{g0} & \frac{1}{2}(\omega_{g1}-\tau_{g1}) & \cdots & \frac{1}{2}(\omega_{gg}-\tau_{gg}) & \frac{1}{2}(\omega_{g1}+\tau_{g1}) & \cdots & \frac{1}{2}(\omega_{gg}+\tau_{gg}) \end{bmatrix}.$$

Let I denote the $(g \times g)$ identity matrix, and proceeding as in Section 6.3, we let

$$M = \begin{bmatrix} 1 & 0 & 0 \\ 0 & I & I \\ 0 & I & -I \end{bmatrix}$$

a $(2g + 1) \times (2g + 1)$ matrix whose inverse is

$$\begin{bmatrix} 1 & 0 & 0 \\ 0 & \frac{1}{2}I & \frac{1}{2}I \\ 0 & \frac{1}{2}I & -\frac{1}{2}I \end{bmatrix}$$

Notice also that

$$M\tilde{\Omega}M = \begin{bmatrix} 2\Omega & 0 \\ 0 & 2\Upsilon \end{bmatrix}.$$

We can then write

$$\theta\begin{bmatrix} e' & \varepsilon' & \varepsilon' \\ e'' & \varepsilon'' & \varepsilon'' \end{bmatrix}(w; \tilde{\Omega}) = \sum_{(m_0,\, m,\, n)\, \in\, \mathbb{Z}\, \times\, \mathbb{Z}^g\, \times\, \mathbb{Z}^g} e^{2\pi i[\quad]}, \tag{6.15}$$

where what fits into the square brackets on the right-hand side is the expression

$$\frac{1}{2}\big[m_0 + \tfrac{1}{2}e',\; {}^t[m + \tfrac{1}{2}\varepsilon'],\; {}^t[n + \tfrac{1}{2}\varepsilon']\big]\tilde{\Omega}\begin{bmatrix} m_0 + \tfrac{1}{2}e' \\ m + \tfrac{1}{2}\varepsilon' \\ n + \tfrac{1}{2}\varepsilon' \end{bmatrix}$$

$$+ \big[m_0 + \tfrac{1}{2}e',\; {}^t[m + \tfrac{1}{2}\varepsilon'],\; {}^t[n + \tfrac{1}{2}\varepsilon']\big]\left[w + \begin{bmatrix} e''/2 \\ \varepsilon''/2 \\ \varepsilon''/2 \end{bmatrix}\right].$$

(Instead of $\theta\begin{bmatrix}e' & \varepsilon' & \varepsilon'\\ \varepsilon'' & \varepsilon'' & \varepsilon''\end{bmatrix}$ we should actually write

$$\theta\begin{bmatrix}e'\\ \varepsilon'\\ \varepsilon'\\ e''\\ \varepsilon''\\ \varepsilon''\end{bmatrix},$$

but this is notationally too cumbersome.)

Now we can do the same trick to the lattice of indices in (6.15) that we have already done several times before, namely, we can sum over

$$L_2 = M^{-1}(\mathbb{Z} \times \mathbb{Z}^g \times \mathbb{Z}^g).$$

In this way we achieve the following expression for (6.15):

$$\sum_{m \in L_2} \exp\left\{2\pi i \left[\tfrac{1}{2}{}^t\left(m + \begin{bmatrix}e'/2\\ \varepsilon'/2\\ 0\end{bmatrix}\right) \begin{bmatrix}2\Omega & 0\\ 0 & 2\Upsilon\end{bmatrix}\left(m + \begin{bmatrix}e'/2\\ \varepsilon'/2\\ 0\end{bmatrix}\right)\right.\right.$$

$$\left.\left. + {}^t\left(m + \begin{bmatrix}e'/2\\ \varepsilon'/2\\ 0\end{bmatrix}\right)\left(Mw + \begin{bmatrix}e''/2\\ \varepsilon''\\ 0\end{bmatrix}\right)\right]\right\},$$

which can be rewritten as

$$\sum_{m \in \mathbb{Z} \times \mathbb{Z}^g \times \mathbb{Z}^g} \sum_{\alpha'} e^{2\pi i[\quad]}, \tag{6.16}$$

where what goes in the square brackets is

$$\frac{1}{2}{}^t\left(m + \begin{bmatrix}e'/2\\ \varepsilon'/2 + \alpha'/2\\ \alpha'/2\end{bmatrix}\right)\begin{bmatrix}2\Omega & 0\\ 0 & 2\Upsilon\end{bmatrix}\left(m + \begin{bmatrix}e'/2\\ \varepsilon'/2 + \alpha'/2\\ \alpha'/2\end{bmatrix}\right)$$

$$+ {}^t\left(m + \begin{bmatrix}e'/2\\ \varepsilon'/2 + \alpha'/2\\ \alpha'/2\end{bmatrix}\right)\left(Mz + \begin{bmatrix}e''/2\\ \varepsilon''\\ 0\end{bmatrix}\right).$$

In the summation α' runs as always over those elements of $\mathbb{Z}^g$ whose entries have only zeros and ones. But the summation (6.16) has a much easier form. We let

$$u = {}^t[z_0, z_1 + z_{g+1}, \ldots, z_g + z_{2g}],$$

$$v = {}^t[z_1 - z_{g+1}, \ldots, z_g - z_{2g}].$$

Then (6.16) is simply

$$\sum_{\alpha'} \theta\begin{bmatrix}e' & \varepsilon'+\alpha'\\ \varepsilon'' & 0\end{bmatrix}(u + \begin{bmatrix}0\\ \varepsilon''\end{bmatrix}; 2\Omega)\eta\begin{bmatrix}\alpha'\\ 0\end{bmatrix}(v; 2\Upsilon)$$

$$= \sum_{\alpha'} (-1)^{(\varepsilon'+\alpha')\cdot\varepsilon''}\theta\begin{bmatrix}e' & \varepsilon'+\alpha'\\ \varepsilon'' & 0\end{bmatrix}(u; 2\Omega)\eta\begin{bmatrix}\alpha'\\ 0\end{bmatrix}(v; 2\Upsilon). \tag{6.17}$$

Here we use the letter η instead of θ to emphasize that this theta function is coming from the period matrix of the Prym variety.

The reason for doing this rather involved computation is that, as the expression (6.17) shows, a relation between theta functions associated with the period matrix Ω of the curve C and theta functions associated with the period matrix Υ associated with the Prym variety $P(\pi)$ is beginning to emerge. But if we choose carefully the values of the various variables, we can actually obtain a situation in which the expression (6.17) is equal to zero. The basic idea for this is as follows. Suppose we take an *odd* theta characteristic L on C. Then

$$\dim H^0(C; \mathcal{O}(L)) \geq 1.$$

But $\pi^*(L)$ is a theta characteristic for $\tilde{C}$ and so

$$\dim H^0(\tilde{C}; \mathcal{O}(\pi^*(L))) \geq 1.$$

But if we can arrange in fact that $\pi^*(L)$ is an *even* theta characteristic, we will obtain

$$\dim H^0(\tilde{C}; \mathcal{O}(\pi^*(L))) \geq 2.$$

If, on top of this, $\theta[\begin{smallmatrix} e' & \varepsilon' & \varepsilon' \\ e'' & \varepsilon'' & \varepsilon'' \end{smallmatrix}](\omega; \tilde{\Omega})$ is the theta function on $J(\tilde{C})$ corresponding to $\pi^*(L)$ under (4.57), (4.58) and the discussion following them, then from (4.60) we can conclude that

$$\theta[\begin{smallmatrix} e' & \varepsilon' & \varepsilon' \\ e'' & \varepsilon'' & \varepsilon'' \end{smallmatrix}]\left(\int_p^q dw; \tilde{\Omega}\right) \equiv 0 \tag{6.18}$$

for all $p, q \in \tilde{C}$. This can in fact all be arranged. The point here is to trace through the many identifications of Chapter Four to see that with respect to the symplectic bases for homology chosen in Section 6.1, the isomorphism (4.56) works out as follows:

$$\text{(points of order 2 on } J(C)) \xleftarrow{\;\cong\;} \Sigma$$

$$I\begin{bmatrix} e''/2 \\ \varepsilon''/2 \end{bmatrix} + \Omega\begin{bmatrix} e'/2 \\ \varepsilon'/2 \end{bmatrix} \longleftarrow \{L\},$$

$$\text{(points of order 2 on } J(\tilde{C})) \xleftarrow{\;\cong\;} \tilde{\Sigma}, \tag{6.19}$$

$$I\begin{bmatrix} 1/2 \\ \varepsilon''/2 \\ \varepsilon''/2 \end{bmatrix} + \tilde{\Omega}\begin{bmatrix} e'/2 \\ \varepsilon'/2 \\ \varepsilon'/2 \end{bmatrix} \longleftarrow \{\pi^*L\}.$$

Thus L is odd whenever

$$e'e'' + \varepsilon'\varepsilon'' \equiv 1 \pmod 2, \tag{6.20}$$

and we can have $\pi^*(L)$ even simply by having

$$e' = 0. \tag{6.21}$$

Even the verification of (6.19) is not too bad. Since its correctness (or not) is invariant under deformation of C, it suffices in fact to check it for the following (degenerate) case, the one where C looks like Figure 6.2. (Compare Figure 6.2 with Figure 6.1.) One then sees that the left-hand component of Figure 6.2 adds nothing to the verification, and what is involved is verifying (6.19) directly, by hand, in the case in which C and so also $\tilde{C}$ are elliptic curves.

So from now on let us assume that (6.20) and (6.21) are satisfied. Equation (6.17) then gives us that

$$0 \equiv \sum_{\alpha'} (-1)^{(\varepsilon' + \alpha') \cdot \varepsilon''} \theta\begin{bmatrix} 0 & \varepsilon' + \alpha' \\ 1 & 0 \end{bmatrix} \left(\int_p^q du; 2\Omega \right) \eta\begin{bmatrix} \alpha' \\ 0 \end{bmatrix} \left(\int_p^q dv; 2\Upsilon \right) \tag{6.22}$$

whenever $\varepsilon' \cdot \varepsilon'' = 1$. To get even a better relation we will sum (6.22) over all values of ε'' such that $\varepsilon' \cdot \varepsilon'' = 1$. To see what comes out we simply must compute

$$\sum_\beta (-1)^{(\varepsilon' + \alpha') \cdot (\varepsilon'' + \beta)},$$

where we sum over all β such that $\beta \cdot \varepsilon_1 = 0$. We obtain

$$\begin{aligned}
2^{g-1} & \quad \text{if } \alpha' = \varepsilon', \\
-2^{g-1} & \quad \text{if } \alpha' = 0, \\
0 & \quad \text{otherwise.}
\end{aligned}$$

So we obtain the relation

$$\theta\begin{bmatrix} 0 & 0 \\ 1 & 0 \end{bmatrix} \left(\int_p^q du; 2\Omega \right) \eta\begin{bmatrix} \varepsilon' \\ 0 \end{bmatrix} \left(\int_p^q dv; 2\Upsilon \right)$$

$$= \theta\begin{bmatrix} 0 & \varepsilon' \\ 1 & 0 \end{bmatrix} \left(\int_p^q du; 2\Omega \right) \eta\begin{bmatrix} 0 \\ 0 \end{bmatrix} \left(\int_p^q dv; 2\Upsilon \right). \tag{6.23}$$

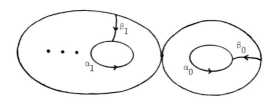

Figure 6.2 A good degeneration of C for checking Prym identities.

But (6.23) must be true for *all* $\varepsilon' \neq 0$ since we can always find an ε'' such that $\varepsilon' \cdot \varepsilon'' = 1$. Our conclusion then is that the vectors

$$\left(\theta\begin{bmatrix} 0 & \alpha' \\ 1 & 0 \end{bmatrix} \left(\int_p^q du; 2\Omega \right) \right)_{\alpha' \in \{0, 1\}^g}$$

and

$$\left(\eta\begin{bmatrix} \alpha' \\ 0 \end{bmatrix} \left(\int_p^q dv; 2\Upsilon \right) \right)_{\alpha' \in \{0, 1\}^g}$$

are proportional for all $p, q \in \tilde{C}$.

6.5 The Proportionality Theorem of Schottky–Jung

The proportionality theorem obtained at the end of the last section is not quite what we want because of the presence of the period matrices 2Ω and 2Υ rather than Ω and Υ. However, we can remedy that with the help of the relation (6.14) which we derived in Section 6.2. We put

$$u_1 = \int du, \qquad v_1 = \int dv,$$

$$u_2 = 0, \qquad v_2 = 0,$$

and we write

$$\theta\begin{bmatrix} 0 & \varepsilon' \\ 0 & \varepsilon'' \end{bmatrix}(u_1 ; \Omega)\theta\begin{bmatrix} 0 & \varepsilon' \\ 1 & \varepsilon'' \end{bmatrix}(u_2; \Omega)$$

$$\underset{(6.14)}{=} \sum_{\alpha'} (-1)^{(\varepsilon' + \alpha') \cdot \varepsilon''} \theta\begin{bmatrix} 0 & \varepsilon' + \alpha' \\ 1 & 0 \end{bmatrix}(u_1' ; 2\Omega)\theta\begin{bmatrix} 0 & \alpha' \\ 1 & 0 \end{bmatrix}(u_2'; 2\Omega)$$

$$- \sum_{\alpha'} (-1)^{(\varepsilon' + \alpha') \cdot \varepsilon''} \theta\begin{bmatrix} 1 & \varepsilon' + \alpha' \\ 1 & 0 \end{bmatrix}(u_1' ; 2\Omega)\theta\begin{bmatrix} 1 & \alpha' \\ 1 & 0 \end{bmatrix}(u_2'; 2\Omega)$$

whereas

$$\theta\begin{bmatrix} 0 & \varepsilon' \\ 1 & \varepsilon'' \end{bmatrix}(u_1 ; \Omega)\theta\begin{bmatrix} 0 & \varepsilon' \\ 0 & \varepsilon'' \end{bmatrix}(u_2; \Omega)$$

$$\underset{(6.14)}{=} \sum_{\alpha'} (-1)^{(\varepsilon' + \alpha') \cdot \varepsilon''} \theta\begin{bmatrix} 0 & \varepsilon' + \alpha' \\ 1 & 0 \end{bmatrix}(u_1' ; 2\Omega)\theta\begin{bmatrix} 0 & \alpha' \\ 1 & 0 \end{bmatrix}(u_2'; 2\Omega)$$

$$+ \sum_{\alpha'} (-1)^{(\varepsilon' + \alpha') \cdot \varepsilon''} \theta\begin{bmatrix} 1 & \varepsilon' + \alpha' \\ 1 & 0 \end{bmatrix}(u_1' ; 2\Omega)\theta\begin{bmatrix} 1 & \alpha' \\ 1 & 0 \end{bmatrix}(u_2'; 2\Omega).$$

The subtlety here is that by (4.39)

$$\theta\begin{bmatrix} 1 & \alpha' \\ -1 & 0 \end{bmatrix}(u_2'; 2\Omega) = -\theta\begin{bmatrix} 1 & \alpha' \\ 1 & 0 \end{bmatrix}(u_2'; 2\Omega),$$

$$\theta\begin{bmatrix} e' & \varepsilon' \\ e'' & 2\varepsilon'' \end{bmatrix}(u'; 2\Omega) = (-1)^{\varepsilon' \cdot \varepsilon''}\theta\begin{bmatrix} e' & \varepsilon' \\ e'' & 0 \end{bmatrix}(u'; 2\Omega).$$

Next we add

$$\theta[^0_0{}^{\varepsilon'}_{\varepsilon''}](u_1;\Omega)\theta[^0_1{}^{\varepsilon'}_{\varepsilon''}](u_2;\Omega) + \theta[^0_1{}^{\varepsilon'}_{\varepsilon''}](u_1;\Omega)\theta[^0_0{}^{\varepsilon'}_{\varepsilon''}](u_2;\Omega)$$

$$= 2\sum_{\alpha'}(-1)^{(\varepsilon'+\alpha')\cdot\varepsilon''}\theta[^0_1{}^{\varepsilon'+\alpha'}_{0}](u_1';2\Omega)\theta[^0_1{}^{\alpha'}_{0}](u_2';2\Omega)$$

$$= (\text{constant})\sum_{\alpha'}(-1)^{(\varepsilon'+\alpha')\cdot\varepsilon''}\eta[^{\varepsilon'+\alpha'}_{0}](v_1';2\Upsilon)\eta[^{\alpha'}_{0}](v_2';2\Upsilon)$$

$$\underset{(6.14)}{=}(\text{constant})\sum_{\alpha'}\eta[^{\varepsilon'}_{\varepsilon''}](v_1;\Upsilon)\eta[^{\varepsilon'}_{\varepsilon''}](v_2;\Upsilon),$$

where the essential point is that the constant is independent of the choice of ε' and ε''. In this way we obtain the theorem of Schottky and Jung that the vectors

$$\left[(\theta[^0_1{}^{\varepsilon'}_{\varepsilon''}])\left(\theta[^0_0{}^{\varepsilon'}_{\varepsilon''}]\left(\int du\right)\right) + (\theta[^0_0{}^{\varepsilon'}_{\varepsilon''}])\left(\theta[^0_1{}^{\varepsilon'}_{\varepsilon}]\left(\int du\right)\right)\right]_{\binom{\varepsilon'}{\varepsilon''}}$$

and

$$\left[(\eta[^{\varepsilon'}_{\varepsilon''}])\cdot\left(\eta[^{\varepsilon'}_{\varepsilon''}]\left(\int dv\right)\right)\right]_{\binom{\varepsilon'}{\varepsilon''}}$$

are proportional, where, for the sake of economy of notation, we suppress Ω, Υ, and $u_2 = v_2 = 0$. In the next section we shall apply this theorem to obtain the Schottky relation.

6.6 The Schottky Relation

We are now ready to derive analytic conditions on the period matrices A which must be satisfied if A is to be the period matrix of a curve. Since it can be shown that the (analytic) set of period matrices of curves of genus g has $(3g - 3)$ dimensions (as long as $g > 1$), the first place to look for a nontrivial analytic relation on the entries of A is in the case $g = 4$. Given the Riemann theta relation and the Schottky–Jung proportionality, the idea of constructing relations is very simple.

We start with an identity between theta-null values, for instance, the identity of Riemann

$$\theta[^0_0]^4 = \theta[^1_0]^4 + \theta[^0_1]^4.$$

Using

$$\eta[^0_0]^4 = \eta[^1_0]^4 + \eta[^0_1]^4$$

and Schottky–Jung proportionality, we obtain

$$\theta[^0_0{}^0_0]^2\theta[^0_1{}^0_0]^2 = \theta[^0_0{}^0_0]^2\theta[^0_1{}^1_0]^2 + \theta[^0_0{}^0_1]^2\theta[^0_1{}^0_1]^2,$$

which must be identically satisfied since the general 2×2 period matrix comes from a curve. Again change θ's to η's and use Schottky–Jung to get

$$\theta\begin{bmatrix}0&0&0\\0&0&0\end{bmatrix}\theta\begin{bmatrix}0&0&0\\1&0&0\end{bmatrix}\theta\begin{bmatrix}0&0&0\\0&1&0\end{bmatrix}\theta\begin{bmatrix}0&0&0\\1&1&0\end{bmatrix} = \theta\begin{bmatrix}0&0&1\\0&0&0\end{bmatrix}\theta\begin{bmatrix}0&0&1\\1&0&0\end{bmatrix}\theta\begin{bmatrix}0&0&1\\0&1&0\end{bmatrix}\theta\begin{bmatrix}0&0&1\\1&1&0\end{bmatrix}$$

$$+ \theta\begin{bmatrix}0&0&0\\0&0&1\end{bmatrix}\theta\begin{bmatrix}0&0&0\\1&0&1\end{bmatrix}\theta\begin{bmatrix}0&0&0\\0&1&1\end{bmatrix}\theta\begin{bmatrix}0&0&0\\1&1&1\end{bmatrix},$$

which again must be satisfied identically. Finally, again changing θ's to η's in this last identity and replacing $\eta\begin{bmatrix}\varepsilon'\\\varepsilon''\end{bmatrix}$ with

$$\left(\theta\begin{bmatrix}0&\varepsilon'\\0&\varepsilon''\end{bmatrix}\theta\begin{bmatrix}0&\varepsilon'\\1&\varepsilon''\end{bmatrix}\right)^{1/2}$$

we obtain a theta identity in dimension 4 which must at least be satisfied for the case in which the period matrix used in building the theta functions comes from a curve (of genus 4).

Finally, it remains to see that the Schottky relation is not simply another relation satisfied by all abelian varieties of dimension 4. For this we will outline an argument given by R. Accola. The beginning point is an elliptic curve E. It is easy to convince ourselves that given any set of eight distinct points

$$p_1, \ldots, p_8$$

on E there exists a Riemann surface

$$h : C \to E \tag{6.24}$$

which is a two-sheeted covering of E branched at the points p_j.

As in (5.5), we build the canonical mapping

$$\varphi : C \to \mathbb{P}_4.$$

Intersections of hyperplanes in $\mathbb{P}_4$ are in one-to-one correspondence with *canonical divisors* on C. Now the double cover (6.24) has a natural involution

$$\iota : C \to C$$

on it so that

$$h \circ \iota = h.$$

This is just "sheet interchange" so clearly $\iota \circ \iota = $ identity map. Now ι induces an involution

$$\iota^* : H^{1,0}(C) \to H^{1,0}(C),$$

so by linear algebra, $H^{1,0}(C)$ decomposes as the direct sum of the subspaces

$$H^{1,0}(C)^+ = \{\omega \in H^{1,0}(C) : \iota^*(\omega) = \omega\},$$
$$H^{1,0}(C)^- = \{\omega \in H^{1,0}(C) : \iota^*(\omega) = -\omega\}.$$

Since the elements of $H^{1,0}(C)^+$ all come from holomorphic differentials on E, we know that dim $H^{1,0}(C)^+ = 1$ and so dim $H^{1,0}(C)^- = 4$.

This last says something important about the canonical mapping $\mathcal{q}$ in this case. It says that there is a point q in $\mathbb{P}_4$ so that under the projection

$$\pi : \mathbb{P}_4 \to \mathbb{P}_3$$

centered at q, $\mathcal{q}(C)$ goes onto an elliptic curve. More precisely we have a commutative diagram

$$
\begin{array}{ccc}
C & \xrightarrow{\;\mathcal{q}\;} & \mathbb{P}_4 \\
\downarrow h & & \downarrow \pi \\
E & \xrightarrow{\;\mathcal{H}\;} & \mathbb{P}_3
\end{array}
\qquad (6.25)
$$

In fact, it can be easily checked that $\mathcal{H}$ embeds E as a curve of degree 4 in $\mathbb{P}_3$.

In what follows, we will only outline the steps. The details are not hard but will make the story a bit too long. Interested readers can fill in the rest with some patience and diligence, or preferably, by conversing with a knowledgeable geometer.

The first thing to notice about (6.25) is that while $\mathbb{P}_3$ has a nine-dimensional set of quadrics (second-degree hypersurfaces), they cut out on $\mathcal{H}(E)$ a seven-dimensional linear system of divisors by the Riemann–Roch theorem (4.9). So $\mathcal{H}(E)$ is contained in a one-dimensional family of quadrics. But the intersection of two quadrics has degree 4 and so does $\mathcal{H}(E)$, so $\mathcal{H}(E)$ is equal to the intersection of any two quadrics of the family. The way we say this is that $\mathcal{H}(E)$ is the *base locus* of the *pencil* of quadrics $\{Q_{(\lambda_0, \lambda_1)}\}_{(\lambda_0, \lambda_1) \in \mathbb{P}_1}$ where $Q_{(\lambda_0, \lambda_1)}$ is given by the equation

$$
\begin{bmatrix} X_0 & \cdots & X_3 \end{bmatrix}
\begin{bmatrix} \lambda_0 q_{00} + \lambda_1 q'_{00} & \cdots & \lambda_0 q_{03} + \lambda_1 q'_{03} \\ \vdots & & \vdots \\ \lambda_0 q_{30} + \lambda_1 q'_{30} & \cdots & \lambda_0 q_{33} + \lambda_1 q'_{33} \end{bmatrix}
\begin{bmatrix} X_0 \\ \vdots \\ X_3 \end{bmatrix} = 0.
$$

$$(6.26)$$

There are four values of $(\lambda_0, \lambda_1) \in \mathbb{P}_1$ such that the determinant of the matrix in (6.26) is equal to zero.

For each of these four values of (λ_0, λ_1) we obtain from (6.26) a quadratic cone in $\mathbb{P}_3$ containing $\mathcal{H}(E)$. One such cone is shown in Figure 6.3. Now each line L on the cone intersects $\mathcal{H}(E)$ twice and if L and L' are two lines on the cone, then

$$(L + L') \cap \mathcal{H}(E)$$

is the intersection of a hyperplane in $\mathbb{P}_3$ with $\mathcal{H}(E)$.

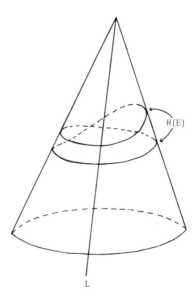

Figure 6.3. One of the four quadratic cones containing $\mathscr{H}(E)$.

What all this means is that any divisor of the form

$$p_1 + p_2 = L \cap \mathscr{H}(E)$$

has the property that

$$2h^{-1}(p_1) + 2h^{-1}(p_2)$$

is a canonical divisor on C, that is, the line bundle on C associated with the divisor

$$h^{-1}(p_1) + h^{-1}(p_2)$$

is a theta characteristic (see Section 4.11). Furthermore, the linear system associated with this divisor has projective dimension at least one since we can move the line L around the cone. If the projective dimension of this linear system were greater than one, then the curve C would have to be hyperelliptic (which it cannot be; if it were, $\mathscr{J}(C)$ would be a rational curve). Thus each of the four theta characteristics obtained in this way is an *even* theta characteristic.

So for each of these four line bundles

$$L^{(1)}, L^{(2)}, L^{(3)}, L^{(4)}$$

we have that

$$\dim H^0(C; L^{(j)}) = 2.$$

Under the correspondence (4.56), the line bundles $L^{(j)}$ correspond to four points of order 2:

$$\begin{bmatrix} \delta(j) \\ \varepsilon(j) \end{bmatrix} \in J(C).$$

So by the Riemann singularities theorem (4.57), there are four even theta functions

$$\theta\begin{bmatrix} \delta(j) \\ \varepsilon(j) \end{bmatrix}(u; \Omega) \tag{6.27}$$

such that

$$\theta\begin{bmatrix} \delta(j) \\ \varepsilon(j) \end{bmatrix}(0; \Omega) = 0.$$

We will skip over the verification that if $\theta\begin{bmatrix} \delta \\ \varepsilon \end{bmatrix}(0; \Omega) = 0$ and $\delta \cdot {}^t\varepsilon \equiv 0$ (mod 2), then the tangent cone to $\theta\begin{bmatrix} \delta \\ \varepsilon \end{bmatrix}(u; \Omega)$ at zero gives a cone as in Figure 6.3. Since there are only four such cones, the four functions (6.27) are the *only* even theta functions which vanish at $u = 0$. There is another point that requires some work (which we will not do). That is, we can choose the symplectic basis for $H_1(C; \mathbb{Z})$ so that

$$\begin{bmatrix} \delta(1) \\ \varepsilon(1) \end{bmatrix} = \begin{bmatrix} 0 & 0 & 0 & 0 & 0 \\ 0 & 0 & 0 & 0 & 0 \end{bmatrix},$$

$$\begin{bmatrix} \delta(2) \\ \varepsilon(2) \end{bmatrix} = \begin{bmatrix} 0 & 0 & 0 & 0 & 0 \\ 1 & 0 & 0 & 0 & 0 \end{bmatrix},$$

$$\begin{bmatrix} \delta(3) \\ \varepsilon(3) \end{bmatrix} = \begin{bmatrix} 0 & 0 & 0 & 0 & 0 \\ 0 & 0 & 0 & 0 & 1 \end{bmatrix},$$

$$\begin{bmatrix} \delta(4) \\ \varepsilon(4) \end{bmatrix} = \begin{bmatrix} 0 & 0 & 0 & 0 & 0 \\ 1 & 0 & 0 & 0 & 1 \end{bmatrix}.$$

Now suppose that we choose the double cover of the curve C in (6.24) for which the Schottky–Jung proportionality

$$\theta\begin{bmatrix} 0 & \delta \\ 0 & \varepsilon \end{bmatrix}\theta\begin{bmatrix} 0 & \delta \\ 1 & \varepsilon \end{bmatrix} = (\text{const})\eta\begin{bmatrix} \delta \\ \varepsilon \end{bmatrix}^2$$

holds. Then for the associated Prym variety we have the identity

$$\eta\begin{bmatrix} 0 & 0 & 0 & 0 \\ 0 & 0 & 0 & 0 \end{bmatrix} = 0 = \eta\begin{bmatrix} 0 & 0 & 0 & 0 \\ 0 & 0 & 0 & 1 \end{bmatrix}. \tag{6.28}$$

So, finally, recall that the Schottky relation has the form

$$\prod_{j,k,l=0,1} \left(\eta\begin{bmatrix} 0 & 0 & 0 & 0 \\ j & k & l & 0 \end{bmatrix}\right)^{1/2} =$$

$$\prod_{j,k,l=0,1} \left(\eta\begin{bmatrix} 0 & 0 & 0 & 1 \\ j & k & l & 0 \end{bmatrix}\right)^{1/2} + \prod_{j,k,l=0,1} \left(\eta\begin{bmatrix} 0 & 0 & 0 & 0 \\ j & k & l & 1 \end{bmatrix}\right)^{1/2}. \tag{6.29}$$

If the relation (6.29) were satisfied for the Prym variety in question, then by (6.28) we could conclude that some

$$\eta\begin{bmatrix} 0 & 0 & 0 & 1 \\ j & k & l & 0 \end{bmatrix}$$

was equal to zero. But, again using Schottky–Jung proportionality, this means that some *other* even

$$\theta\!\left[{}^{\delta}_{\varepsilon}\right]\!(u;\Omega)$$

vanishes at $u = 0$ for the (Jacobian of the) curve C, that is, some $\theta\!\left[{}^{\delta}_{\varepsilon}\right]\!(u;\Omega)$ not in the list (6.27). But we have already mentioned that for most C of the type occurring in (24), *no other* even theta function vanishes at $u = 0$. Thus we can conclude that the relation (6.29) is not satisfied for all 4×4 period matrices, which is what we set out to do.

This is not, of course, the complete story even for 4×4 period matrices. Since Schottky's time there has been considerable progress on this problem of characterizing period matrices which come from curves, but the question is still far from being completely answered.

References

We include here only those references cited frequently.

1. GRIFFITHS, P., and HARRIS, J., *Principles of Algebraic Geometry*. New York: John Wiley and Sons, 1978.
2. GUNNING, R. C., and ROSSI, H., *Analytic Functions of Several Complex Variables*. Englewood Cliffs, New Jersey: Prentice-Hall, Inc., 1965.
3. GUNNING, R. C., *Lectures on Riemann Surfaces*. Princeton, New Jersey: Princeton University Press, 1966.
4. HIRZEBRUCH, F., *Topological Methods in Algebraic Geometry*. New York: Springer-Verlag, 1966.
5. LANG, S., *Linear Algebra*. Reading, Massachusetts: Addison-Wesley Publishing Co., 1971.
6. MUMFORD, D., *Abelian Varieties*. Tata Institute of Fundamental Research, Bombay: Oxford University Press, 1970.
7. O'NEILL, B., *Elementary Differential Geometry*. New York: Academic Press, 1966.
8. SERRE, J.-P., *A Course in Arithmetic*. New York: Springer-Verlag, 1973.
9. SPRINGER, G., *Introduction to Riemann Surfaces*. Reading, Massachusetts: Addison-Wesley Publishing Co., 1957.
10. STEENROD, N., *The Topology of Fibre Bundles*. Princeton, New Jersey: Princeton University Press, 1951.
11. VAN DER WAERDEN, B. L., *Algebra*. New York: Fredrick Ungar Publishing Co., 1970.

Index